S0-ADF-367

LITERATURE AND HISTORY
OF
AVIATION

LITERATURE AND HISTORY
OF
AVIATION

Advisory Editor: JAMES GILBERT

NO ECHO IN THE SKY

Harald Penrose

ARNO PRESS

A NEW YORK TIMES COMPANY

TL
540
P43
A3
1972

Reprint Edition 1972 by Arno Press Inc.

LC# 78-169433
ISBN 0-405-03776-7

Literature and History of Aviation
ISBN for complete set: 0-405-03789-9
See last pages of this volume for titles.

Manufactured in the United States of America

NO ECHO IN THE SKY

272923

Harald Penrose

NO ECHO IN THE SKY

With headpieces by
KEITH SHACKLETON

CASSELL · LONDON

CASSELL & COMPANY LTD
35 Red Lion Square
London, W.C.1

and at

210 Queen Street, Melbourne; 26/30 Clarence Street, Sydney;
24 Wyndham Street, Auckland; 1068 Broadview Avenue,
Toronto 6; P.O. Box 275, Cape Town; P.O. Box 11190,
Johannesburg; Haroon Chambers, South Napier Road,
Karachi; 13/14 Ajmeri Gate Extension, New Delhi 1; 15
Graham Road, Ballard Estate, Bombay 1; 17 Chittaranjan
Avenue, Calcutta 13; P.O. Box 23, Colombo; Macdonald
House, Orchard Road, Singapore 9; Avenida 9 de Julho
1138 Sao Paulo; Galeria Güemes, Escritorio 454/59 Florida
165, Buenos Aires; Marne 5b, Mexico 5, D.F.; Sanshin
Building, 6 Kanda Mitoschiro-cho, Chiyoda-ku, Tokyo; 25
rue Henri Barbusse, Paris 5e; 25 Ny Strandvej, Espergaerde,
Copenhagen; Beulingstraat 2, Amsterdam-C; Bederstrasse
51, Zürich 2

© *Harald Penrose*, 1958
First published 1958

UNWIN BROTHERS LIMITED
WOKING AND LONDON
F 158

FOR

PETER and JIMMY, MICHAEL and TED

and all those others who flung their brief days
against the weighted scales of time

I am indebted to

EDWARD LANCHBERY

for his perceptive advice and sympathetic editing.

HARALD PENROSE

CONTENTS

Give me the wings, magician! So their tune
Mix with the silver trumpets of the moon,
And, beyond music mounting, clean outrun
The golden diapason of the sun.
There is a secret that the birds are learning
Where the long lanes in heaven have a turning
And no man yet has followed: therefore these
Laugh hauntingly across our usual seas.
I'll not be mocked by curlews in the sky;
Give me the wings, magician, or I die.

HUMBERT WOLFE

CHAPTER I

The Wind in the Wires

The metallic drone of stuttering engines ebbed and flowed upon the calm air. Box-kite biplanes, their white wings veined with shadowed ribs, swayed through the evening sky. Slender monoplanes, like darting dragonflies, skated nervous circuits of the field, and settled upon the turf.

In that year, 1912, the music of flight was a compelling inspiration. It held a magic that made even the unattainable seem within reach of those insubstantial wings of wood and fabric, fashioned into rigidity yet ethereal as gossamer drifting in the sunlight. Founded upon methods of trial and error, these bold ventures in scientific thought were built in the breathless enthusiasm of first discovery, and the revelation that flight offered a new and startling beauty hitherto the prerogative of the gods in the heavens. To many the quest was imperative: fly on rapturous wing towards unknown adventure; scale unconquered heights; reach for the stars! The gateway of new worlds was open.

Graham Gilmour, Gordon Bell, F. P. Raynham, Pierre Verrier, Salmet, Pégoud, B. C. Hucks, Gustav Hamel, and many another I watched soaring into the thrilling skies. I remember the upsurge of incredulous excitement when Raynham stopped the propeller of his new Avro over Hendon aerodrome on a cloud-flecked February day when people generally believed that devoid of motive power, an aeroplane must drop from the skies to its doom. Yet here was Raynham demonstrating perfect control over that tiny cruciform shape, high above the bright edge of a cumulus cloud. Intently we watched the aeroplane descend, growing larger and more detailed until we could read the name AVRO painted in bold black letters along the white fuselage.

'He'll do one of his famous tail-down glides in a moment,' said someone. 'Look!'

Ten feet off the ground the Avro changed its steeply tilted glide to a flatter path. The tail dropped level. The machine skimmed a couple of feet above the turf, and to everybody's delight, instead of running his wheels on to the ground, with tail high in the manner of the Blériots and Morane monoplanes, Raynham brought the tail lower and lower until wheels and tail-skid were almost touching the grass. The speed fell away and the machine settled lightly in a perfect three-pointer.

I joined in the applause at this novel trick which was to become the correct method of landing until the advent of tricycle undercarriages thirty to forty years later.

All too soon, it seemed that these frail romantic wings had been but prelude to the holocaust of war greater and more terrible than man had ever experienced. Suddenly aeroplanes were forced to a maturity of lethal purpose. The savage purr of fighters with powerful rotaries and the ominous beat of twin-bombers gave a deeper echo to the skies. Suspense and despair joined battle above the ruined cornfields of the Flanders Plain. Over the trench-scarred mud, single-handed combats swayed between knights who used no armour but their skill: Rhodes-Moorhouse, Ball, Guynemer, McCudden, Mickie Mannock; yes, and Immelman and Richthofen as well.

Then they too had gone—winging into the great vault of oblivion with the same finality that my hero, Gustav Hamel, whose photograph held place of honour in my bedroom, vanished whilst flying his racing Morane-Saulnier above the Channel in the spring of 1914. Indeed, most of the pioneers of the first decade had finished with their flying, and handed to yesterday's youthful spectators the opportunity of experiencing the same enlightenment and pleasure that enchant all who attain the skies.

As a schoolboy, I welcomed the dawn of that eagerly awaited day of my initiation in 1919 with both apprehension and promise of fulfilment. The morning sunshine was filled with the chorus of birds. I listened, dreaming of the world of wings and what might be. Today was the test—the culmination of all my yesterdays; maybe the beginning of a different end, for my secret hope was one day to become an airman.

The minutes of the morning dragged into hours, but at last it was happening. I was cycling to the flying field. At intervals the slow swell and fade of a rotary engine droned in the distance. In the last dusty mile I heard an aeroplane take off but its flight was hidden by dense, towering elms.

I pedalled faster, and arrived in time to see a biplane landing with a sigh of whistling wires and a spasmodic growl from the engine. Its black nose shone in the sunlight. The propeller was a golden blur. The racy stagger of the narrow square-tipped wings was emphasized by twin rows of brightly varnished struts leaning forward into the wind. Against the dark paint of the slim fuselage, ending with comma-like rudder, a final flourish was given by the yellow undercarriage skid fixed between the wire-spoked wheels.

As the machine sailed across the whitethorn hedge the wings rocked to a burst of engine, and I caught a glimpse of the pilot, his hair streaming in the wind and goggles lifted to his forehead. Accompanied by a hollow rumbling from the under-carriage the khaki Avro landed on the rough pasture, and the valves clicked like knitting-needles as the engine stopped.

I propped my cycle against a fence and breathlessly entered the field. Beneath a limp windsock, beside a battered car and a stack of red Shell petrol cans, stood the Avro in all its spindly elegance. The air was redolent with romance, blended from crushed grass, burned engine oil, and the pear-drop perfume of newly doped wings.

All my hopes were pinned on the next few minutes. Yet I hung back, for my heart thumped with excited anxiety. I must not fail myself and find I was afraid—for if I were, then I should never soar on wings of wonder, higher and higher, parting the rolling clouds, climbing the mists of space, spanning the world. So might I dream, for how could I guess that in the topmost heavens, before which lands and seas unroll in a vast remoteness, might be found new truths. Instead, my awareness was little more than a blind groping towards the thrill of the unknown, and the unformulated quest of adventure and fulfilment.

I moved to the aeroplane. The parchment-like fabric of the fuselage towered above me. I set my right foot in the lower foothole, and groped with my left for the other half-way up. Anxiously I swung myself into the padded hole of the cockpit. Then I was sitting on the plywood seat, fumbling at the safety belt and strangely isolated from the outside world that had lost its reality. The pilot smiled reassuringly, and I heard his voice: 'Petrol on, suck in.'

The propeller was pulled over, first one blade and then the other. Beneath its cowl the unseen rotary engine turned with a hollow bubbling noise and the valves clanked with sharp precision.

'Contact!'

I gripped the seat hard. The engine spluttered into a shattering noise that shook the machine, and the slipstream battered my head. A cloud of blue smoke swept past. It was pungently sweet. The throb of cylinders brought the aeroplane to life and filled it with a sense of power. The pilot 'blipped' the switch on the top of the control stick, and the wings swayed to the bursts of engine. He waved a signal with his hand.

Like a half-bemused creature stirring itself, the Avro 504 lurched and moved, drawn forward by a propeller that had raced its identity into a misty disc. A harsh roar enfolded me. The wind turned to a hurricane rippling the fabric sides and beating my face. I gasped for breath. Now for it! We were moving—gathering speed.

Everything except the aeroplane became blurred. Between struts and bracing wires, sky and trees and field blended into a pastel haze of indistinguishable green and blue and sunlight. Bumping and swaying we rushed into the wild press of air, drawing from the great wind a strength that tautened the wings and bracing. The jolting eased, and the machine was lifting—lifting into a divine floating over hedges, fields and tree-tops that had acquired an unusual, rounded perfection and glinted with unexpected light as they drifted beneath. Buoyant as a boat, we sailed above airy depths, the wind of our motion sustaining the wing-surfaces with such easy power that flight lost all sense of peril.

I stared through the cascading slipstream, beyond the lower wing, at a slowly moving backcloth of ordered trees and hedges. Those were the barriers that a few seconds ago had bounded my world, shutting me in like the walls of a garden. But now I knew that they could be surmounted, and stretching far away beyond them was the promise of unexplored horizons.

As far back as I could remember I had been interested in natural things and conscious of the stirring of a vague sense of affinity with the wilderness world. In that moment, airborne at 500 feet, I was suddenly closer to an understanding of the inexplicable. I knew that my life, and the wings on which I was poised, and the skies and the earth, were integrated in some profound design.

I looked about me with greater composure. A deep serenity encompassed the morning world. The spirit of the countryside seemed to reach up to me, hinting of revelations that might one day be told. The roaring of the engine was forgotten, as though deep silence reigned.

A gale might rage outside, but to lean forward beneath the brief protection of the few inches of windscreen was like sitting on a wind-blown hill high above land. The perfume of the earth in spring mingled with the exhaust of burnt oil. I could feel the very texture of the air, feel its lift and strength.

Abruptly the green countryside tilted, rising swiftly above the port wing-tips. I grasped a strut and leaned away from the earth to the starboard wings etched nakedly against the sky. The pilot sat, unconcerned, staring ahead, as the horizon slid past the rounded top of the engine cowl. With relief I realized that the aeroplane was merely turning, and my sense of equilibrium came tingling back. Yet the instant of disquiet whispered a threat that the art of piloting was more complicated than level flight suggested. Was this easy balancing— this gauging of the degree of bank to match each turn—a supremely difficult art? The wind hurled away my questioning.

The wings sank, and we floated in the springtime calm over a countryside more beautifully intimate than I had ever dreamed. Glittering sunlight dappled trees and fields, and transformed the blossoming hedges to bright foam. Secluded thatches, gable ends and mellowed walls, showed half-hidden among the trees, each cottage and garden a miniature universe, harmonious with the earth that bore them. Everywhere the land bore the mark of husbandry and tillage: here, the sparse, tender green of young wheat; there, a hillside grazed to velvet smoothness, its lower slopes scored with straight furrows of newly ploughed loam.

The roads men trudged were dusty strands of brown, shadowed and half-hidden by hedge and tree, yet obvious of purpose like the fine lines of path and track crossing many a field. A stream flashed; a tall steeple speared towards our wings; fields carpeted with flowers slid by, a stile, a gate, a farm; cattle, the embodiment of peaceful ease, trees and more trees. Except for its dream-like quality seen through the gulf of tangible air beneath our wings, the perspective had lost its strangeness. I seem to have known it all my life, and the gale on which we rested was familiar as any cliff-top wind.

Again the earth began to lift above the wing-tip, and I could not stop the involuntary contraction of my muscles. A vast silence filled the universe—an eerie silence, of which the engine rhythm was no longer a part. Wind moaned in the wires and sighed against the windmilling propeller. We were plunging steeply towards the earth. Again mind and body tensed with disquiet. Not an engine failure?

The far vista vanished. Trees loomed into giants of jeopardy, each meadow walled with green armour. The ground rushed up, the aeroplane lurched, and a sideways wind blew so violently upon my face that it made me gulp for breath. Another jerk, and the aeroplane was miraculously level, skimming low across a hedge. A staccato burst of engine and the nose lifted steeply. We were floating close to the turf.

The blur of grass slowed into a sharp focus that picked out each blade individually, and we brushed the earth with the slightest bump. Another half-hearted burst from the engine drew us rumbling across the meadow, our speed decreasing rapidly. With a glance at the wing-tip the pilot ruddered sharply, turning the Avro on its tracks to a dying mutter from the engine.

The propeller stopped and the blustering wind was gone. A curious, half-familiar ticking came from the stilled engine as it cooled. A small group of people stood by the hedge, and nearby were piled red cans, and a drooping windsock. Almost incredulously I realized that this was the field from which we had started. The landing had been perfectly normal.

Everything was still. I sat unmoving, smelling the mingled oil and acetone, wondering if it was really I who had flown. The pilot, in his cockpit beneath the tall struts of the centre-section ahead, turned an oil-streaked face and smiled, waiting for me to clamber out so that he could follow.

The silence vanished. Time moved again. I stepped down to the firm earth beside the khaki fuselage. A bird was singing in the hedge: many birds were singing. Their song was jubilant, rising triumphantly to the skies—and through the song I hear a far voice calling, calling, promising most wondrous things.

CHAPTER II

Give Me the Wings

Presently the quest led to an unexpected world of abstract and applied science, where symbols became a system of logic explaining basic natural laws on which the elements of every mechanism and structure are founded. Electrons, calculus, stress and strain, erg, modulus, moment, thrust: these were the words of my new vocabulary. Sandwiched between the course of aeronautical engineering at London University were two six-monthly periods of apprenticeship at aircraft factories; and the way was paved for me to join in 1926 the Westland Aircraft Works under the direction of Robert Bruce, an inspiring and lovable pioneer designer. He imbued me with his spirit of enthusiasm and high endeavour. Were we not blazing the trail that might lead to argosies of commerce and the new conception of a brotherhood of man linked by the close association

19

that world air-travel could give? Maybe it was the vocation that enthralled—the romance, the pursuit of perfection, the sense of satisfaction at the solution of each new problem—and therefore our eyes were blind to the tools of war we forged. But for me the dream of flying ever beckoned until it became a longing that coloured all my hours: a fulfilment I must achieve. Soon opportunity came.

The litany of flight was murmured down the speaking tube of an open-cockpit training biplane into my ears: simple, economical words which opened secrets and gave confidence to break the fetters of the earth. Absorbed in the confusing impressions of initial flight—the blur of movement, tilting sky and earth, engulfing roar, gulps for breath, tightening of muscles at each bump—I was at first startled to hear sudden speech in the solitude.

'Put your right hand on the stick and feet on the rudder-bar.'

The wind tore at my goggles, and pressed like cold fingers on my mouth. I groped for the controls.

'O.K.? . . . Don't jam them too hard. Right! I want you to follow the movements . . . watch the horizon.'

Expectantly I peered round the edge of the narrow windscreen. Far beyond the beating rocker arms of the three protruding engine cylinders a great sequence of tree-bordered fields spread remoter and smaller until their individuality was lost beneath the clear, encircling line that separated land from sky.

'It's quite simple . . . I raise the nose by pulling the stick back—so. . . . To drop it, I ease forward—like this. . . . If a wing is down I pull it up with a firm movement of the stick in the opposite direction—so.'

Forgetful that the instuctor is in the front cockpit and cannot see me, I stiffly nod my helmeted head.

'To steer, take a point on the horizon—that hill will do. If the nose swings to the left, like this, press the right foot and it comes back—so. But if you want to turn to the left, press with the left foot: to the right with the right. . . . Got the idea?'

'Yes.'

'Good! Now keep the machine in straight and level flight. . . .
You've got her.'

The magic of all first times filled that moment. With new
understanding I perceived that the aeroplane was sentient,
eager and responsively vibrant with life. Through the taut
cables of ailerons, elevator, and rudder, she became linked with
my body. Thought flowed to fingers and feet guiding her flight.

Flight! This was freedom, untrammelled, bearing me
buoyantly through the air. With what confidence the wings
leaned upon the wind, rocking gently, lifting and falling with
the easy motion of a ship.

'Keep her steady. . . . You're letting the nose weave up and
down.'

Hurriedly my gaze reverted fixedly ahead, and I pressed
forward a little on the top of the stick. The nose dropped
unexpectedly far. I gave what seemed an almost imperceptible
tug backwards, and the horizon sank far below the engine
cowl. A firm insistent pressure against my palm warned that
the instructor was correcting the errant flight-path with his
dual control.

'You have it,' he said after a moment.

I grip the stick too fiercely and jam the rudder bar. Mind and
muscle are needlessly tensed. There is no time to see the little
world of men and trees and fields, nor the mock formality of
toy-like houses. All is a patterned blur of green and brown
beneath an upturned bowl of mist-blue sky, at whose rim I
stare to orientate myself in space.

But in a little while the nose ceased wavering, and I discovered that aeroplanes fly best when left alone, restrained
only by the briefest, lightest touch.

Time was forgotten in the absorption of this delightful task,
holding her level, wings parallel with the edge of the world;
steering steadily at a distant mark. All too soon the instructor's
voice:

'Good enough for a first effort. I'll take her.'

Then the soft whistle of the glide; the school aerodrome—busy with silver-winged biplanes—coming into sight ahead, growing larger, nearer, closer. The tilted earth and a sudden pressure on one's head as the aeroplane side-slipped in to land; a smooth contact; the jolting of wheels on rough turf, the surprising stillness of the world as the engine died. The initiation was over.

Before I could venture alone there was an all-important fence that had to be surmounted. In those sharp, thin wings of old lay a menacing trap. If one flew too slowly, the airflow would break down with vicious suddenness and drop the machine into an uncontrollable spin. An over-ruddered turn, a gust of wind on the slow glides of these old, lightly laden machines, or a moment's abstraction and failure to hold the right speed, could easily prove fatal.

So came the final test. Flying circumspectly level in the high, sunny air was like walking on a cliff-top path, subconsciously aware of the danger of falling down the abyss, yet caring not because of the splendid view.

The instructor's black helmet was impassive in the forward cockpit. I took a deep breath. Close the throttle, ease back the stick. The needle of the air speed indicator dropped slowly back. I pulled up the 'cheesecutter' lever of the elevator trimmer, easing the hand load required to prevent the nose from dropping. The sky seemed waiting. The aeroplane was leaning back upon the air. When I tentatively moved the control column sideways the ailerons barely rocked the wings. I pulled the stick a fraction further: the machine gave a disquieting lurch. I pushed on full rudder. The nose swung sharply down, and the horizon tilted upwards dragging the fields askew.

With a steady whirl the aeroplane spun and the land was lost in a rotating blur. Only by the dizzy flick of the sun as it passed on each revolution could I count the turns: two, three, four, five. I centralized rudder and stick in succession. The aeroplane swung on for a half a turn more, and then the horizon sank to its usual place in the cowling as I pulled at elevators made unnaturally heavy by the speed.

I opened the throttle and the spinning disc of the propeller blades flashed in the sunlight. The altimeter showed that I had lost a mere 1,500 feet. My reflection showed in the instrument's face. I ruddered to hide the glare, and felt glad the test was over.

'Do another,' said the voice in my earphones.

I obeyed. The instructor's head remained motionless, but I felt the pressure of his hands and feet taking over the controls.

'I've got her.'

Our return to the aerodrome was as near as anyone can get to the poetry of movement. In sweeping curves of one turn after another, switch-backing from climb to dive and thence to a long, shallow glide, the aeroplane danced into final approach, side-slipping and swish-tailing, before floating like thistledown on to the summer turf.

The instructor climbed out casually.

'Off you go,' he said. 'If you feel confident after the first, you can do a few more landings.'

Flight succeeded flight, and assurance grew, and my climbs reached higher. Rising in great circles, the sun would throw the shadow of my head alternately on each lower wing, and cast on the drum-tight fabric of their surface the silhouette of struts and wires. As the world of green sank into remoteness, a universe of space, brilliant with light, became my long-dreamed heritage. This sense of breathless discovery was like opening a book of wisdom written in a strange, entrancing language of sunlit, cloud-patterned hills, and valleys that seemed to be imbued with the mystery of life as though some languorous spirit dwelt in their folds. So glorious yet poignant was the earth in her unguarded beauty that I was too enthralled to understand a tithe of what I saw; yet there, in form and colour, glossed with the pattern of the present, was the shadowed story of the past, and the key to mystery after mystery.

Each airborne hour found a new discovery. What if a generation already had enjoyed these revelations with identical delight? It was sufficient that my eyes had never previously lit on them, for me to experience the same thrill as those pioneer explorers. With every flight the fetters of the earth became unloosed; yet, with each return the mystery was lost and the echo of the revelation made unreal. I was an earth man, dreaming he had flown. With feet firmly planted on the ground I gazed into the skies and watched with envy the accomplished flight of birds.

Gradually from the medley of impressions, England took shape. Lifting my goggles I would peer through the slipstream to view the landscape with clearer, naked vision. It was like looking from a boat at bright sand and seaweed-covered rocks shining clear through watery depths.

With map and magnetic compass I steered from town to town, comparing the map colours of graded brown and green with the living landmarks of hill and wood, threaded by roadway, rail and stream. Pictorial representation and the land itself became synonyms of one another. In my hand was the geographer's image of Cornwall or Kent. There, far below, the map had sprung to life with form, solidity and colour. Rejecting the custom of man-invented names for town and hill and river, I saw the world with something of the vision of a bird, and recognized where I was by physical features so individual that labels were needed only to define them. The bones of the land lay contoured into hill, valley and plain, cloaked with a prolific garment of life, or naked, gaunt and uncaring.

No matter how many times I might fly above a landscape grown familiar, some new comprehension might dawn and hint at revelation beyond the physical act of sight. A hundred times I would look down upon the same scene, and then unaccountably, the next time the view would be different, transformed by instinct and a vision of truth before an inner eye.

Ah yes! This was flight, slow and intimate, passing with

lingering gaze in a long, sweet dawdling on the honey-scented wind. Even the most powerful of those early aeroplanes moved with joyous grace upon the velvet textured air; and the machines that I flew were fragile, shining creatures of fabric stretched on varnished frames of wood clipped together with thin metal fittings, tied with bracing wires, and held with little bolts.

Lightly laden wings responded to the slightest turbulence with the sensitivity of a soaring bird. In spite of the sacrificial toll of those who wagered with their lives and failed on desperate occasions, this was a fair dawning through which the spirit of flight matured. No sense of gamble stirred our thoughts for long. Ours was a magic carpet where we sat on cushions high enough to see beyond the obstruction of the cockpit's padded rim. The seat was wicker, or riveted from metal, devoid of adjustment and rarely comfortable. But who wanted comfort? Leather coated, helmeted and goggled, gauntleted and sheepskin booted, we sat with slipstream rushing past the windscreen and pummelling head and shoulders, while icy draughts eddied through the laced sides of drumming fabric. Upon frozen lips and cheeks the air pressed pure and cold as moonbeams, but each breath was headier than the grape. This was our youth, our heritage. The cup of freedom was filled to the brim with expectation and the hazard of adventure, and a sense of splendid power. Inwardly jubilant, we sat in our cockpit waiting for the undiscovered horizons to unfold.

CHAPTER III

I'll Not be Mocked

To fly my first relatively powerful aeroplane after slow school machines was to perceive the mighty forces that were barely under man's control. It was a lesson in respect, tempered with the thrill of adventuring into a widening sphere of exploration and achievement.

The Bristol fighter, a two-seater biplane to which I was promoted in 1927, impressed and awed me. It had a wing span of nearly forty feet and was powered by a magnificent star-shaped, nine-cylinder, radial engine. I had already sat many times in its cockpit with hopeful anticipation, familiarizing myself with the controls and the upright seat posture, and becoming accustomed to the boxed view between the wings and over the broad nose.

Helmet and goggles under my arm, white silk scarf knotted

and tucked into overalls, I settled into the comfortable wicker seat in an atmosphere of amyl-acetate blended with rich oil. An extra cushion, thrown up by the ground crew, gave me the right height, and another wedged behind my back brought the rudder bar within easy reach. I pulled on my helmet, and buckled the broad safety belt round my waist. Checking the simple fuel system from the diagram on the dashboard I turned on the main cock, and glanced through the avenue of cross-braced wires and raking struts to watch the arrival of the grotesque Hucks starter vehicle. Its overhead chain-drive transmission ended in a spring-loaded toggle, mating with a claw on the aeroplane's propeller hub. A mechanic coupled the starter, and the fuselage quivered on its elastic-mounted undercarriage at the impact. I unscrewed the Ki-gas primer and doped the engine.

'Contact!'

With a wheeze and a shake the Hucks locked into gear and turned the propeller. I switched on, and whirled the stumpy handle of the starting magneto, setting off a grating roar and a puff of smoke.

The engine ticked over, stirring the aeroplane into a life that flooded my consciousness and submerged the ties of earth. It was the prelude to a different incarnation.

Through the bright blue of the slowly flailing blades I watched the starter truck back away. I signalled for the engine run-up. Two mechanics sprawled across the fuselage in front of the fin and weighted down the tail, their indifference to what was one of life's great moments for me lost in the grim endeavour of withstanding the slipstream.

The engine roared and rattled through 1,200 to 1,400 r.p.m., and then a brief burst of exhilarating thunder at 1,600 r.p.m. I held it there, checking switches and pressures before throttling back to rumbling quietness. Having set the tail-plane incidence to take-off I pushed the goggles over my eyes, and waved away the chocks. Now was the moment.

A man at the wing-tip struggled against the leading edge,

as I applied full rudder and aileron and with careful bursts of engine swung the machine into wind. The turn told me something of her temperament. We taxied over the cropped grass and stopped, preparing myself for the moment like an actor awaiting his cue. The spark of anticipation grew to a flame. I looked cautiously round the skies and saw no other aeroplanes.

I opened the throttle. The grass streamed faster beneath the lower wing, the wind changed to the familiar gale, the aeroplane pulled away from the mark on which I had aligned the nose. More in desperation than belief that the machine was within power of control, I jammed on full rudder and checked the swing with such ease that confidence flowed back. In the manner taught with the school trainers I pressed lightly forward on the control column and brought up the tail to flying position. The aeroplane accelerated as though on rails.

'Gently! She's a woman . . . treat her firmly but gently,' came the remembered injunction of my instructor.

Through the triangulated pattern of wires, the hedge and trees of the windward boundary hurried closer, growing large and seemingly unavoidable. But in that moment we ceased to bound across the turf almost out of hand and, gathering speed, lifted with sweet assurance. Responsive and steady the fighter climbed steadily. Trees lost their hostility and dropped beneath with a last glittering flourish of their topmost leaves. The horizon spread wider. A great triumph filled me with exultant happiness as from the safety of my newest wings I watched the world turn Lilliput and more remote. Soon it was no more than a carpet beneath the echoing sky. The aeroplane and I were united in single solitude.

My basic flying experience was founded more on this machine than any other. After a dozen hours of practice of circuits and landings and familiarity with normal flight, a more extensive exploration of flying characteristics began. Slowly we ventured towards the imagined but unknown: a safe enough path, for the Bristol was one of those aeroplanes hall-marked with the universal approbation pilots gave to the manner of its flight. Its responses particularly pleased. They were born of those imponderable qualities of stability and control which are difficult to analyse and impossible to repeat identically in some later conceived design. That is where the test pilot exhibits his scientific art in trying to marry physical and psychological impressions with the technicians' mathematics of movements and reactions. The object is to ensure that whatever the speed of flight, the load needed to operate controls shall be within a pilot's effortless power, and produce conventional responses within limits of his mental ease. All this the fighter had, and other less definable attributes, which added to her perfection and the confidence with which we flew. So secure was my kingdom that I forgot how slender were the four ash longerons forming the fuselage girder in which I sat.

In an atmosphere of growing confidence I took the Bristol on a proving flight on a day of perfect summer. When we were high enough I let the aeroplane fly level, and tried to feel determined whilst my mind and senses fought as they did when I was a small boy making my first dive from the highest board. For days now there had been an inescapable compulsion to test myself, and the act required was to perform a loop. Officially that stage of instruction was not in the curriculum, so I learned the method from a few oblique questions and earnest study of the R.A.F. training manual.

Tentatively I pushed down the nose. The speed crept up, and I felt more and more convinced that the racking strain imposed upon the machine must presently break it. Controls grew heavier. Moving them was like levering great weights. I levelled off quickly. There was no parachute to reassure me.

Once more I let her dive until the rigging wailed a high unbearable note, and, fearful of straining the aeroplane, I pulled timidly back on the control. Up and round swept the fighter, her dark nose lifting and lifting, the earth dropping away, the horizon obliterated. Because I did not realize that by looking outwards past the wing-tip, instead of over the nose, the side horizon could still be used as datum, I lost all sense of attitude. A frenzied glance at the air-speed indicator showed that the speed was near the stall. I had pulled too gently on the stick. Instinctively I pushed on hard rudder with my foot, and in a breath-taking Immelman the aeroplane cartwheeled sideways into a dive.

It was so perfect an experience that I did five or six more, getting the feel of the machine as well as confidence in strange, steep attitudes so different from the mild departures from straight and level to which I had been trained.

The wings of exhilaration were beating for me now. Far below was the serene green world. Down there were rules and laws; but up here was a sunlit empire of ecstatic phantasm, where the bondage of inessentials was broken and one's heart beat in tune with the universe.

So on to the sunlit depths of space I leaned: free of deliberation: pressing the control: waiting. . . . Down, down, down into the gulf. Yet subconsciously my mind was guardian, noting the A.S.I. needle, moving steadily round to 120 . . . 130 . . . 140. The air was dense and strong, straining the structure with tension, but now I was prepared for it. I was no longer concerned when the controls became immensely heavy. I gulped at the oily, solid air hurtling past the windscreen. Instinct told me that the aeroplane was ready, settled for a jump. I pulled back against the elevator load.

Up, up, up went the nose. The blue sky grew dark as the machine reared vertically and the blur of the propeller slowed. The load on the controls lifted. I pulled the stick fully back. For an interminable time the machine seemed suspended stationary in space—then it wallowed further backwards, and I was hanging by my belt, feet dangling free from the strapless rudder bar. I clung to the stick to save myself from falling out. The propeller turned so slowly that I could see each blade making its own circuit. The sky was ominously quiet. A voice within me promised that if I were spared this once, I would never be such a fool again.

Unexpectedly it was over. The nose flopped further down. I dropped with a jolt on to the cushions, and the aeroplane was sweeping into a wire-screaming controllable dive, completing the loop.

For ten minutes I flew cautiously around, my pulses still tingling. Then with sudden determination I nosed down once more, holding speed and attitude briefly before pulling gently but firmly on the stick. The fighter sailed round in a perfect loop. It was intoxicating and easy, and I accomplished ten more loops before I landed.

For the last of these I flew thirty miles to the Westland aerodrome. The result of this unauthorized shoot-up was unexpected. The company's directors took note of the fact that I could handle a Service aeroplane, and allowed me to combine my job as works manager of their civil aviation department with demonstration flying.

In this I received great help and encouragement from Westland's chief—and often, only—test pilot, Louis Paget, a colourful character who had won the D.F.C. in the First World War, and sported a monocle and loud check suit as his props and costume on the stage of the world. Paget guided me into the flying of the firm's experimental and military types, and I was sent to South America to demonstrate the Wapiti general purpose biplane.

The day after my return, Louis Paget spun into the ground in a Widgeon. He survived with fractures of the leg and thigh but never flew again, and I found myself a wholetime test pilot.

On many a day thereafter the sky became the setting for aerial ballet in three dimensions. In flowing curves from one circle to another in every plane from zenith to nadir, I danced my adagios and pirouettes. Sometimes I felt as though I painted these motions with the sweeping brush-strokes of a Michelangelo upon the canvas of the universe. My medium was light; the colours from youth's palette, with no thought that they could ever dim to softer harmonies of pleasure.

Yet it was more than that—for an uplifting inspiration welled through my senses and compelled these aerial curves and sweeping motions. To send the earth, with a twist of the wrist and touch of the foot, swaying and turning in a spacious arabesque, or to hurl it dropping and spinning from sight, then to snatch it from blue oblivion, was to experience a sense of mastery with inner calm after long striving for perfection. It was not the feeling of power that counted, but the confidence of experience in testing myself.

At other times, when the poetry of flight was made subservient to more deliberate science, I would try with spring-balance and stop-watch to measure in terms of force and time the secrets of stability and control behaviour which made my Bristol fighter so perfect a machine. In the simplicity of such a beginning I did not foresee that in the course of years I should make hundreds of similar recordings in a long sequence of aeroplanes. Yet so it was to be.

In England, the U.S.A., France, Germany, throughout the world, men laboured to make aeroplanes in many different forms. In quiet Wessex where we planned and played, the creations that I flew were diverse and often strange. Some were bat-like tail-less types; others had rotating wings; monoplanes emerged whilst biplanes held high favour; wood and fabric developed into metal. Bombs grew bigger, and the fire-power of guns increased; yet so engrossing to the senses is the flying of an aeroplane that there was rarely occasion to reflect where all this evolution must lead. Each flight was creation of the moment. Each new, untried machine demanded the same explorative caution and offered the same subconscious vehicle of escape to the magic casements of fulfilment that no earth-bound man could know.

Deep as the stillness on high hills was the serenity in the high places above the earth. Like whispered sweetness beneath the stars was the calm reassurance given by the inner triumph of self-conquest after the battle with each new experience.

CHAPTER IV

Saga of a Widgeon

A light aeroplane called the Widgeon became my greatest tutor. Setting the trend for Westland design which was to culminate in the Lysander, the Widgeon was a high-wing monoplane with a span of thirty-six feet and a top speed of just over one hundred miles an hour. With this machine I skimmed the countryside finding more unhurried view than any other aeroplane could give. From 1928 we flew together learning secret after secret of the countryside and sky. I was a youth when I took her for her maiden flight on that far day of her first gay summer when they wheeled her, new and glistening, from the workshop. I was twenty years older, still dreaming myself unchanged, when I started her on the last, which ended a few seconds later in the disaster of leaping flames.

No other aeroplane was ever so well-suited for aerial observation, for there was practically nothing to obstruct the

c

downward field of vision. The 'parasol' wing was placed on struts high above the body so that the pilot, seated far aft, had superb views, not only below, but upward as well. Yet there was more in the arrangement than unimpeded view. An exceptional degree of stability at the stall enabled the Widgeon to be flown with confidence at speeds slow enough to match the flight of many birds.

I liked to let this little aeroplane whisper through the air with engine almost closed, my gaze ranging far across both sky and ground. If the wind blew strongly it was possible to remain almost stationary over places of interest; yet, when the voyage of discovery was over, only a touch on the throttle was needed to bring her racing home at 100 m.p.h.

What tremendous discoveries we made—what wonders we saw! So many of those important and ever-remembered first occasions belonged to the Widgeon.

Again and again I watched with enchantment England take form from the coloured map in my hand. The buff and green paper picture became soaring brown hills rising from vales patched with meadows and patterned with hedgerow trees. But the map could not show the immensity of space as we sailed far and wide through the skies of youth, nor could it hint at the miracle of seeing this island with a comprehension that a hundred years of travelling on foot could never give. With each successive flight the individual characteristics of every county became more recognizable, until they were patently related to the geological structure of the land.

I began to see the countryside not merely as a cameo of exquisite beauty, but as the expression of the gigantic forces of expansion and contraction unleashed on the world as it grew from elemental form. Its later history was there to read as well: old boundaries of the sea, newer incursions, dried river valleys; turf-disguised mounds and marks of early man; Norman roads and Saxon paths; castles of conquerors uncertain of their hold, side by side with gracious dwellings of a more peaceful age, and the staring council houses of today's planned state. I saw too the grimness of industrial centres set like scars on the fair face of England, their smoke and fog covering a third of the countryside upon which the sun once shone unveiled through uncontaminated air.

I learned also the trick of height which dwarfs a river to a brook, or a mountain to a molehill: I knew then why the gods were omnipotent. Yet I found that other days could turn the mountain into fearsome walls, low-capped with cloud, where the Widgeon flew ensnared and only faith could find the valley of escape. And in the evening mist the estuary, which from 5,000 feet could be covered by my hand, was from a few feet high, a widespread trap of ugly waves, waiting for me to lose the vague horizon and sink within its shroud.

There were days, too, when the Widgeon took me into the world of cloud. Patiently we climbed, penetrated the overcast and became engulfed in the dark swirling vapour, travelling in a blind eternity that masked all sense of equilibrium. Suddenly the mist would grow luminous, become white, and in another instant we would burst through the top of the cloud into a world of brilliant light. As the little aeroplane climbed higher the clouds stretched wider and wider below, like an endless snowfield glittering under the arch of heaven. There, it was always as though we hung motionless, transfixed by the incandescent disc of sun on which no eye dare rest, so blinding was its majesty. The troubled world of people receded beyond thought and I felt reborn in that sublimity of space, oblivious of the truth that winged man was no more than a moth fluttering round the beacon lights.

Sometimes there were flights in the dusk, with the lights of villages and towns springing one by one in company with the early stars, until it seemed the countryside was scattered with glittering jewels. Sometimes we flew by moonlight—the hills and valleys grown strange and mysterious in the blanched light, and the sea a dusky silver framing the dark loom of the shore.

On many an occasion the Widgeon had borne me along that same coastline by daylight, until I knew the south like the back of my hand, and the east and west passably well. Over every small harbour I had flown, circling while I watched the way of men with ships. In storm and calm, sun and driving rain, at ebb and flood, I had seen the endless changes of the ocean's face, until at last I felt I understood a little of its power and inappeasable emptiness.

We watched, too, the passage of the seasons, slowly flooding the length of these isles with green growth to attain the high tide of summer, and then ebbing to the regeneration of winter days. Twenty times we saw those black months of waiting and brooding on buds anew, until the countryside became suffused with the soft glow of richer colour that presages spring. Then the chess-board of ploughed soil would change texture as the earth grew hidden by thrusting shoots of new grass and tender wheat. Trees would froth into lacy leaf, and in the course of a week or even a day, the land would be transformed into eager, green, proliferous life.

From the Widgeon I would see white blossom sweep the hedgerows, and the fields grow starred with flowers. Yet to fly a little north could mean passing back from the wave of spring to find the buds still waiting to unfold. But only for a little while; soon the whole land would be verdant and fulfilled . . . and as spring drove ever northwards, summer followed close.

From the air we saw infinitely delicate transitions as summer grew mature. Tree and hedge grew darker, velvet meadows changed to tumbled swathes. Presently there was golden grain. How quickly the flush of spring was forgotten when squares of plough showed once more among the fields of stubble and sere green; and autumn made the landscape glow with bronze, painting the earth with strange dim magnificence, like the tarnished chattels of a once noble house. With the first touch of winter the coloured trappings disappeared, the last leaves fell and trees were transformed into black tracery. Yet as the aeroplane skimmed over the topmost branches I could see that the bare twigs held already in their swelling tips the promise of yet another reincarnation.

So the months and years came and went, with questing and fulfilment, sadness and happiness. All that time the Widgeon carried me on tranquil wings, revealing new beauty in every hour, trying to let me understand at least a little of the mystery, the cohesion, the marvel and the eternity of nature and the cycle of life.

Often, of course, we forgot the passing of time and played, for we were light-hearted as the birds. The breeze whispered exciting wordless promises that set veins tingling. The controls were delicate as finger-tips which stroked and read the air. For a little while the Widgeon ceased to be a man-made machine, and I knew only that I had wings. Across the sun-filled sky we glided like skaters over ice, cutting voluptuous curves and airy figures, dropping a few feet above the aerodrome, sinuously twisting and sliding a handspan over the turf. Then up and up, breasting the air, towering like a peregrine after its stoop.

Ah yes! Many times we had seen the peregrine at play do that—hurtling in its pounce like a searing blue arrow, only to rocket upwards and away. At other times we found the bird soaring at ease, throwing circle after lazy circle on a thermal lift, a mile above the world. . . . Not only the peregrine but every bird of the skyways we watched as well. They were kinsmen—but with an older prerogative than mine. Theirs was an understanding and an unthinking art, perfected by the evolution of a million years.

Turning, crabbing, hanging at the stall, the Widgeon drifted far across the countryside whilst I studied the manner of a bird casually encountered. Or perhaps we might seek them in their haunts: herons on the network waterways of Sedgemoor; buzzards mewling above wooded Devon hills; this family and that of wildfowl sheltering in their hundred thousand on the winter floods; grey geese with singing wings. Sometimes we searched for strange birds such as harriers, knowing that success would be rare, but that the prize of discovery would be all the more valuable.

37

Always it was like that. Even the last flights we made together were in their way a triumph—for now it was high summer and we played with the birds as of old.

It had been the Widgeon's unluckiest year. For months she had been laid up—initially to change her ancient engine for one of newer vintage, and then because she stumbled in her next flight, losing a wheel and breaking her mahogany propeller. When at last the Widgeon flew again, it was as though a pretty moth had emerged from the chrysalis of long hibernation, for she had been re-built and painted handsomely with silver and soft sky-blue.

Over a brilliant landscape, no whit less miraculous and beautiful than those of her twenty other summers, the Widgeon flew with the carefree wending of a butterfly attracted by the gold heart of every flower. I watched the world slide slowly past, bringing each minute a score of new delights. From the few hundred feet at which we flew, the countryside was intimate as a garden, yet with a broad and beckoning vista drawn clearly like a map. The old enchantment gripped me, and I counted myself lucky to be of a century which at last could look at the world from the skies and see secrets known to a hundred million years of birds, but never before revealed to man.

The air was buoyant as the sea, but smoother than the stillest pond. Expanded by the earth's heat it shimmered upwards, bearing on its steady up-currents many insects and birds. On restless wings swallows flashed steel-blue as they tilted in the sunlight. Above them the dark crescent forms of swifts carved across the sky as they hawked the gauzy *diptera*. But though in numbers far exceeding other birds, the swifts and swallows were not alone in exploiting the heated air. Kestrels hovered on the breeze, silhouetted fox-red against the green of turf and shrub, their wing-tips blue. Where the woods were wildest on the hillsides, buzzards lightly circled on upspread wings, lifting steadily until they were high specks. Away to the north, above the water meadows, a heron sailed, dark wing-pinions flexed to pointed tips which gave no beat in half a mile. With all these birds, as well as the circling rooks, the Widgeon played on each of these last few flights; but mostly she flew contented and alone, wrapped in her dreams, happy to press and lift against the warm, calm air of yet another summer.

No new discoveries, no first encounters, no sudden revelation, no high adventure: only those three flights of tranquillity over the fairest country of all the world.

All too soon it was the end. For the last time those silver wings had borne me through the limpid air, and I came sliding down in sighing curves to land lightly upon the new-mown turf. I switched off the engine. Silently she stood there, facing the little breeze, while I flew another, newer aeroplane, forty times as powerful and five times more swift. The Widgeon waited, perhaps in jealousy, fearing I had given my heart to a new love, not dreaming that none can compare with the old. And everywhere birds were singing and calling her, for had we not sought and followed them everywhere?

Presently I was ready to go home. I swung her propeller. The engine gave a sudden, all too vigorous roar. Before the man standing by her switches could cut the power she was moving and running away.

I flung myself at the cockpit and clambered on to the steps. But the ecstasy of the summer was nerving her, and high above was the wild cry of the mewling buzzard. She flung me off, and pilotless skimmed the ground, for the last time pressing herself against the breeze as she lifted.

So little a freedom this, so short a flight, so quick the end to aspiration. Twenty yards, fifty, she flew by herself—rocking a little, turning uncertainly. Then she swung tighter, and, only a few feet off the ground, went crashing full-tilt into the side of a shed. The roar of her engine died. Silence a moment, broken only by the song of a thrush—and then, under a pall of black smoke a great yellow flame leaped up, higher and higher. Her silver wings vanished, her gay, blue body became a stark black skeleton. Within three minutes there was only smoking wreckage, a last flickering flame, and a burned-out hole in the shed.

A swallow flashed past, turning with airy grace from wing to wing. High and far away, I heard again the faint cry of the buzzard. From the copse at the end of the aerodrome drifted the muted talk of rooks. I looked high into the sky, and saw the swifts still soaring on crescent wings.

The smoke faded and rose more slowly. But from the flames, from the ashes, no Phoenix arose—only ghosts of memories, and the haunting knowledge that there would still be other days and other flights, using this aeroplane and that, but none so loved as the little Widgeon who had taught me that flight was more than self-revelation, and greater than the materialization of a happiness akin to ecstasy. She had let me hear the echo of the very universe. To fly was a form of worship, a discrimination of mighty patterns transcending human life, instilling in me the desire to share the vision which the Widgeon had first shown me when she and I were young.

I turned to her again. Above the burnt-out wreckage a last slowly rising column of thin smoke hung insubstantially against the depth of sky. When it had gone a longer chapter of my life was closed. I breathed farewell—to you, my Widgeon, and you and you and you bound inescapably with the pageant of her years—and yet I knew that this was not the end. The future breathed promise that there would be other days and other years, more and yet more, to gather to my heart.

CHAPTER V

Winged Pegasus

So many times in summer and winter during those early days,
I sat waiting, half-dreaming, half-watching, whilst the biplane
I was testing climbed steadily upwards. Three-quarters of an
hour usually saw the ceiling reached, and the reserve of power
so spent that barely enough was left for the labouring propeller
to pull the aeroplane precariously forward. Meanwhile my
heart beat deeply to compensate the lungs for the attenuated
frozen air they breathed. Sometimes my temples throbbed or
grey shades slid like ghosts across the limits of consciousness.
What did it matter? Such flights were exercises in self-
abnegation and youth's conquest of the frailty of life.

41

On the occasions when oxygen was used the cold remained a bitter enemy; but one December day in 1932 neither frost-bite nor the slow, excruciating return of warmth to frozen hands and feet could make a penalty of flight. Although I sat in an open cockpit my clothes, including face-mask and goggles lenses, were luxuriously heated with electric elements like nerves netting all my body. My face was protected by grease and a larger windscreen than usual took the sting out of the icy coldness of the whipping slipstream. I had only to sit but slightly crouched to gain the shelter of the rudimentary roof and side-panels.

On the giant alcohol thermometer strapped to the rear member of the interplane struts, the reading was already ten degrees centigrade below zero. Linked by an oxygen tube to the matrix of the aeroplane, I turned my encumbered head to the altimeter and found we had reached four miles. At the altitude we must presently achieve, if so many hopes were to be realized, the bitter coldness of the air would be greater than any Arctic solitude.

This sun-bright emptiness high above the endless snow-field of the clouds was the exclusive province of lightly laden single-seaters with a powerful engine. Yet the prediction of the aerodynamicists was that our specially stripped, supercharged two-seater Houston-Westland would climb a mile higher than the ceiling of the Bulldog fighters.

On the result of this climb depended the achievement of another milestone in man's endeavour. Those who had worked these last three months on the project were confident of success, although in the beginning before all the factors had been fully assessed, it had seemed an ambitious gamble. If the range and ceiling of the aeroplane, laden down with heavy cameras, proved adequate, the first attempt in the world was to be made to fly over the unconquered peak of Everest. Calculations showed that if a Pegasus radial engine, which had achieved the single-seater world-altitude record, were fitted to the best two-seater general-purpose aeroplane of the day, a ceiling of a marginal 4,000 feet above the mountain might be achieved. This assumed a standard steady atmosphere, but no one knew how great the down-currents at the mountain peak might be.

Steadily climbing on a long, straight slant we simulated the flight path to an imaginary Everest, 160 miles beyond the starting-point. Above the clouds the air was frozen into stillness without wind or thermal current to ripple the smoothness of the flight. We seemed suspended in immobility. Only the creeping hand of the altimeter, dawdling with the dragging minutes, signalled that we climbed. I felt—but my senses, so long noise-conditioned, no longer heard—the harsh thunder of the engine. Instead it was the far faint music from beyond the stars which registered in the mind, so that I was half-god and half-robot, dreaming my thoughts, whilst methodically recording the instruments and measuring with stop-watch the time for each 1,000 feet ascent.

When twenty-four minutes had gone and 25,000 feet showed on the altimeter we had reached the highest I ever had been. With the most critical part of the climb looming ahead, I stirred in my parachute straps to ease my cramped body, and turned up the flow of oxygen. Breathe steadily; don't be flustered by the strangeness; take it easily, warned my thoughts.

Below, the floor of cloud was pierced here and there with ice-blue depths that led back to the world. I could see a patch of England's southern shore edging a hammered metallic sea. Below the starboard wing-tips the diamond shape of a mist-blue Isle of Wight thrust like a wedge into the glazed waters of the Channel.

In the unexplored silence of sun-blazed space I stared through the void of distance. Time, with majestic beat, changed from a measure made by man and became the throb of the universe echoing the ghost reflections of all that had flown into the abyss of the past, and all that would happen in the aeons to come. Here was the still, serene voice of the

Absolute that turned to naught man-made ideologies, religions and philosophies. The vastness lapped like a great tide upon my wings and slowly bore me higher. Looking upward I saw the blue-black hem of the curtain, which, could I but lift its corner, would reveal the radiant inexplicable.

My spirit was disembodied, watching me fly through a golden mist of light. The sky was blue-black with infinite distance, reaching out into the endlessness of space where unseen stars might make some vaster universe. Beneath that blue-black canopy of terrible emptiness, the Houston-Westland was a microscopic, frozen-winged, silver creature poised defenceless and frail in solitude, lonelier than the farthest star. Automatically I turned the oxygen regulator further to maximum flow, and my senses cleared. I was almost six miles above the world, but no recollection whispered that this was the attainment of Everest's height. No vision came of the mountain, glittering icy white in the illusive distance. My goal was simply a figure on an instrument, the purpose of its achievement forgotten in the striving to success.

A lifetime back I had waved away the chocks and opened up the engine. Yet, since I pressed the stop-watch as the aeroplane lifted from the grassy aerodrome, only thirty-eight minutes had passed. I looked away from the beauty of this solitude and turned to the lightly quivering needles of the instrument panel.

Another row of figures was added to the pad strapped to my knee: height, time, air temperature, engine r.p.m. and boost, oil readings, oxygen flow. My head clearer, I turned back the life-giving regulator a trifle to conserve the oxygen.

Breathe the cold oxygen stream slowly and deeply: sit unstraining: feel the heat of the clothes. Learn the lesson that comfort aids staying power and morale. Comfort for the pilots of the next decade must mean protective cabins and warmth, and then in its turn pressurization. Like me those pilots to come will gaze far below at the bluish whiteness of cloud strata, and see the remote edge of the world blending its icy distance with the pale rim of the bleached blue sky. Looking upwards they will find the blue turning deeper and darker until in the zenith the blue is dusted thickly with black. But however high they climb, however far they fly, there, above the topmost cloud, the inescapable blazing whiteness of the sun, overwhelming in its splendour, will dominate all heaven and earth.

While thought played with its eternities the climb grew slower and more laboured. In the attenuated air the super-chargers could no longer feed the engine with enough oxygen to burn the fuel efficiently. The haze of the fanning propeller blades changed. As height grew greater, engine speed visibly decreased. I began to will the aeroplane to make its climb, holding with careful precision the forward speed appropriate to such height. I sat there, still and waiting, while the seconds lengthened to hold a score of fantastic thoughts.

High in this solitude that gave no sense of loneliness, a realm of vision beyond mortality opened. No inessentials spun their captive web. Life bloomed like a peerless flower upon the airy summit of my world of light. With no recollection that the earth lay shrouded in shadows beneath the floor of cloud, there was no comprehension that mankind lived in bitterness and grief, in happiness and hope, and loved and died and was re-born throughout the million years. Mine was the illusion of immortality, and a conviction of vast power reaching from outer space—and all the time there was still the exaltation of that far faint music of the spheres.

A hundred stars were scintillating on the aluminium instrument dash. The aeroplane had turned slightly while I dreamed, so that the sun's blaze struck across my shoulder on to the polished metal. Gently I banked back on to course, making the first deliberate control movement throughout the hour. In the intense atmospheric cold, the light alloy wing frame-work and the steel control cables had contracted at different rates, giving slight slackness to the wires and making less effective the response of an aeroplane whose stability had been noticeably diminished by the thinness of the atmosphere. This and the remoteness of the tilting horizon made balance feel precarious. I was glad when we returned to level flight.

The climb crept on again. Fourteen more minutes became fourteen days of staring at the hypnotic flicker of the propeller, while heart-beats drowned in waves of noise that receded to a mute drone, then loomed large again. The altimeter paused, moved, hesitated. Thirty-five thousand feet changed to thirty-six, and to thirty-seven thousand. The thermometer was showing minus 65 degrees. Here at last, where the temperature fell no more, was the edge of the stratosphere and the threshold to the mysteries of space of which astronomer, poet and scientist had so long dreamed.

I tilted the wings at maximum effective incidence to the attenuated airstream, but the engine's vitality was gone. Gone, too, was the propeller's grip. In ten minutes of exhausted clutching at the sky only 500 feet were won. The aeroplane rocked and sank, lifted wearily and sank again spent by its endeavour. Turning the oxygen to maximum flow I wrote a last row of figures—then pulled back the throttle.

A profound and startling silence leaped from the solitude, frozen with overwhelming desolation. Then the bracing wires swelled a shrill protesting cry as the aeroplane began to gather speed for its long descent. For a little while I could relax, glad that the test was almost over. In a few more hours when I sat by my evening fireside the flight would have an unreal, dream-like quality.

Even now, beyond the downward tilt of the nose, blue sky had given place to the homely, familiar aspect of clouded earth fringed by the Channel seas. Far below projected Beachy Head; and separated by the grey waste of water was the French coast picked out by the low sun of the winter afternoon.

In simulating the flight path planned for the mountain attack, I had been restricted to the theoretical time estimated for reaching the crest. Sufficient fuel had been calculated to give a short period of reconnaissance and photography over the peak, leaving a comfortable margin for the return flight since most of it would be descending with throttled engine. I checked the petrol content's gauge. Forty-six gallons had been used in ninety minutes: barely the figure estimated. I gave a routine glance at the spy-hole prismatic indicator to check that fuel was being pumped from the main to the gravity tank which fed the engine. In place of the dark colour of flowing petrol was the emptiness of silver. The solitary pump, unduplicated for economy of weight, had failed. The tank could be nearly drained. At any moment the engine might cut for lack of fuel.

I struggled to remember how long ago I had last checked the feed and found it normal. Ten minutes? Twenty? More? Impossible to remember in that rarefied atmosphere.

Home in that instant became far away. I swung the machine on to the reciprocal course for base. When the compass steadied I saw that everywhere the cloud stretched white and unbroken, the gaps filled in, closing the downward corridors of escape.

After the first swift appreciation of danger the mind worked disembodied, with cold deliberation weighing the prospect, assessing the relative risk, scheming out a complex pattern of 'if' and 'if'. In later years a calmly impersonal voice would have reached across the ether giving the compass course to safety, guiding the descent through cloud at the right speed and place to make an accurate pre-determined landfall. But now there was no such help. Radio aids were still a dream. Tomorrow would be won only if today, and in this hour, no mistakes were made, and fortune smiled.

But whilst I thought and hoped and wondered, the engine note insidiously changed and the pulse of life that vibrated throughout the machine became uncertain. It was happening. The engine was stopping. Five seconds later the power faded and vanished. Only the slipstream strove to spin the broad wooden propeller against the compression of the lifeless engine. Presently that, too, gave up the struggle and the propeller

47

stopped altogether. I pushed down the nose to overcome the increased drag of that carved block of wood, and turned off the ignition and fuel to lessen the danger in the event of a crash. I remembered the electrics of the heated suit and turned off the master-switch too. Cold began to creep through my clothes.

Thirty-five thousand feet and more than a hundred miles from home, with all the grassy aerodromes of England hidden beneath the mantle of cloud. What should one do? Where should one steer? True, there should be 10,000 feet of clear air beneath the stratus that would allow twelve minutes in which to select a field big enough for landing—but the risk of damaging a machine on which such hope depended hung its threat over my head. I took off a gauntlet and blew warmth on my freezing hand.

Unhurrying, the silver biplane glided towards the unknown end. Only the dropping altimeter hand warned of the approaching crisis.

From the illusion of safety in my shielded fabric and metal shell I looked upon the vast expanse of cloud, and turned the aeroplane to study the southern escarpment below which a long edge of coast untidily protruded. Somewhere not far within that cloud fringe the aerodromes of Shoreham, Tangmere and Hamble were dotted among the fields. I had to make up my mind which I should try to find when I plunged into the clouds and broke through the last vapours to find the dull shadowed land.

My instinct knew the answer. I turned the aeroplane with deliberation. In whispering silence the aeroplane dropped, its structure inanimate, its heart-throb lost. Steadily, in a long, straight line, it glided towards the middle distance of my view, where the shores of Spithead and Solent converged on Southampton Water and the hidden notch of the Hamble river, by whose edge the aerodrome named after it lay unseen. All my thoughts were concentrated on it. The high, dark summit of sky that had so enthralled me, the frontiers of mystery, the vista of unending time, the quest and the attainment had become no more than drifts of speculation long vanished in the stream of thought that is the consciousness of life. So I sat there, unmoving, waiting for the seconds to pass that would bring the cloud floor close at last to my wheels.

At first height gave illusion that the route to escape was within control, but when a bare half-mile separated aeroplane and cloud, the feeling changed. I was committed and encompassed. The perspective shortened. The south edge of the cloud layer was lost to view. From palest tints of blue the surface changed to rugged undulations. Steamy fingers groped at the wings, solidified and blotted out the light and freedom of the high skies in vaporous darkness.

No blind flying instruments: only compass and bubble cross-level for reference; so press the rudder lightly in opposition to the needle's movement, and let the aeroplane's lateral and longitudinal stability bring her through. It was an act of faith contradicting all the body's senses—yet no greater than trust in the complex gyro indications of later days. A bare two minutes saw the thinning mists of lowest cloud, the dawning light, and sudden emergence above a sober landscape, so unreal and forbidding after the brilliance above that it might have been some new, strange planet. I tore off mask and goggles, and searched the dull panorama for landmarks.

There was no need to worry. Between the stationary propeller and the stark line of lower wing, spread the channels of Chichester, Langston and Portsmouth harbours, and beyond them a little way in the smoke-hazed distance shone the silver line of Southampton Water with the Hamble entering near its mouth. The wind-song in the wires sounded a more confident note. The aerodrome was easily within reach.

Glimpses of grey cruisers and gun-heavy battleships; open country; the blue gloom of the Wight across water away to the left, and twin aerodromes between coast and wings. Why had I forgotten Gosport and Lee? But I must not change my

mind now. Senses and judgement are conditioned to acceptance of landing at Hamble. Hold air speed steady, machine
in perfect trim. The natural termination of the glide path angle
extends far beyond the aerodrome, so there will be height on
arrival for a half-circle into wind.

Nearer and lower, the view spreading wide: Southampton
smoky beyond the right wing-tip: yachts at their anchorage
below Hamble village: black and silver school aeroplanes
trundling across the aerodrome: a great ship moving down
Southampton Water—and I still have nearly 2,000 feet in hand.

In the corner of the aerodrome the yellow windsock points
limply to a gentle south-westerly breeze. Even a down-wind
landing would be safe, for, brakeless though she is, the run of
my lightly laden aeroplane would be scarcely 200 yards. But
this is no emergency. Why land down-wind when there is
height in hand to make a normal circuit and choose my time
to turn in for the approach? . . . Gently the aeroplane canted,
gliding left-handed along the water's edge, past Fawley
Beacon, past the Hamble entrance, sweeping wide of Warsash,
turning towards the long row of anchored yachts—and 800
feet still left.

Now was the time to complete the turn. Reducing speed
slightly the aeroplane came slanting across the river, but I
kept her a little high to give a safety margin. The aerodrome
was ahead. With lightest pressure on stick and opposite rudder
the machine side-slipped over the hedge, and the wings were
dwarfed against the widening spread of grass. A smooth movement of the controls, and the aeroplane sailed level for a last
few seconds, then kissed the rustling grass blades and swayed
lightly across the ground, settling and slowing to stop 200 yards
from the tarmac apron in front of the hangars.

I undid the safety belt, took off my helmet, and unclipped
the electric elements. On my face the cold breath of a winter
breeze brought no memory of the bitter coldness of the heights.
I was earthbound in the world of men, and did not pause to
think that we had already lifted the veil over Everest.

CHAPTER VI

Gift of Days

A great wind blew; but from a cloudless sky the sun painted rich colours on the purple Suffolk heaths and tawny stubble fields in the late summer of 1934. There was a glory in the wind-swept air, for this wild, heather-covered aerodrome where I waited was in the other country of my youth. Flying its clear skies I could look down and see my past pictured by familiar shapes of wood and stream; a quiet creek clustered with anchored yachts; a white road bordered by tall elms winding towards a mill upon a hill; tall spires, quiet villages, and in the distance a great estuary filled with ships, and the sea sparkling beyond. All that countryside and coast re-kindled hidden thought and memory. In a few more minutes the curtain would rise upon the stage, and with absorbing pleasure I would see its friendly setting once again.

Beneath the wide wings of the Westland P.V.7, the biggest high-wing monoplane that had yet been made, I buckled on a borrowed parachute, and with subconscious caution tightened its over-long leg straps in case the wildly improbable need to use it should happen on that flight. I leaned to the wind while the engine was being tested. Its echoing thunder shook the peaty ground. Four mechanics sprawled on the tail, their clothes fluttering and clinging to their bodies in the fierce slipstream. The run-up ended, and the huge wooden propeller was flickering slowly round once more, shaking the bull-nose of the aeroplane under the erratic impulse of throttled running. The mechanic jumped down and I took his place in the triplex-panelled cabin built level with the wing. No longer was it necessary to crane and peer between silhouettes of nose and mainplane. This high position behind the engine gave an incomparably finer view than from any cockpit placed behind the wings. Indeed this torpedo-carrying monoplane represented a great aerodynamic advance on the biplanes which so long had been in favour. We all hoped and expected it to prove so superior to its competitors that it would be adopted by the Royal Air Force.

Locked in the novelty of a glass-house cabin I was insulated from the outside world, which assumed the reflected unreality of an image in a mirror. As I sat there in the sunlit warmth the transfiguration from earthbound mortality began. When the aeroplane moved forward a primitive thrill of expectation stirred within me.

The turbulent atmosphere struck viciously as the big mono-plane began to rise. The wings rocked under the assault, but checked immediately to the pressure of the controls. We rose swiftly above the brown and purple aerodrome, hovering with only a little forward motion as the aeroplane pressed against the gale, and pitching and rocking as height was gained. At 2,000 feet we still were over the edge of the aero-drome, creeping slowly forward although the indicated speed showed 70 m.p.h.

At eye level a broad expanse of wing stretched on either side, braced to the fuselage from half-way along the span by wide pairs of downward sloping struts. If I lifted slightly in my seat the arched top surface revealed its silver doped canvas sucked upwards into tight little arcs between each rib. It was equally easy to look backward along the fuselage and view the tall fin and massive tail. The strength and solidity of the aeroplane's structure gave tremendous confidence. This was no frail beauty but a superbly engineered example of aeronautical art.

The Suffolk landscape spread wider, brilliant with sunlight, as we climbed. Ten miles away the low coastline looked no more of a barrier to the sea than the green edge of a lily leaf is to the ripples of a pond. In the still, warm cabin I had forgotten the strong wind conditions until I saw the foam-flecked ranks of waves. Only by looking steeply down at the earth was it possible to appreciate the strength that could sway the tree-tops like rippling corn. A slow procession of hedgerow and field drifted under the massive legs and bulging wheels of the undercarriage.

I drew in a breath of air pungent with acetone and oil. Gone the days of buffeting wind, cascading icily over a windscreen far too small. My leather helmet and thick flying suit hung obsolescent in a clothes locker. I sat bare-headed and wearing summer-light slacks and jacket. The silk gloves on my hands were only for protection from the sharp edges of screws and nuts that seemed inseparable from every aeroplane. The discomfort of flight was gone; but so, too, was the reality of being airborne. Flight had surrendered the exuberance a bird might feel at the velvet lash of air around its wings. We were boxed with artificiality, and something of that first fine careless rapture was missing.

Yet the spell remained no less enthralling; indeed the wonder and the revelation grew. Each advancement in design brought the attainment of still wider vistas unknown before this age, dream though the philosophers might since man was capable of thought. That was my fortune, fulfilment, and my delight—for there below lay the world revealed for the airman to read, its story growing clearer with every turn of the page.

The big P.V.7 climbed steadily, moving only slowly forward as it breasted a wind grown stronger with height. The altimeter showed 10,000 feet, but beneath the tail the brown rectangle of aerodrome was still visible. I made a routine check of the instruments and pencilled their readings in the ten columns on my knee pad. With five minutes to go before the next set need to be taken, I mentally began to plan and prepare for the test. The green and gold landscape blurred into a background for my thoughts. I had become part of the mechanism: the control which set the sequence in response to the automatic computer and found subconscious answers, but was human enough in prescience, for the essence of the test was to confirm there was no danger of the abnormal loading affecting stability when the machine was dived to the limit of its speed. In the meantime the climb and engine data must be recorded. Another set of figures was added to the pad.

The rate of climb diminished as the air thinned and starved the low-altitude engine of power. I, too, was without oxygen, but 14,000 feet would be high enough to make the preliminary dives, and low enough to avoid concern for the heart and breathing.

As the needle of the altimeter reached its mark I levelled the aeroplane and moved the hand-wheel at my side, adjusting the tailplane so that the machine was trimmed for level flight. A last check was made. I turned in my seat to look under the wings at the newly devised drag-flaps and made sure they were closed. A glance upwards confirmed that the sliding cantilever roof was firmly shut. Either flap or canopy out of position might produce a marked change in controllability at high speed. But all was well. This was the moment—and it was always the same.

I braced against the shoulder straps and, lightly pressing forward on the control column, slowly pushed the nose down. The horizon lifted higher and higher, growing like a wall. The airspeed needle swept round faster and faster. The fixed pitch propeller spun swifty in the increasing wind, and I throttled back a little to avoid exceeding the maximum permissible revs. My mind was intently occupied with the ever-changing message of the instruments, but it subconsciously appreciated that all was well: 3,000 feet, 4,000 feet, 5,000 feet were lost and the speed was up to 210 m.p.h., not far short of the maximum permissible.

I flattened out, bringing the aeroplane sweeping up through the shallow arc, and then, aided by the initial impetus of its speed, climbed quickly to 15,000 feet. I spared a moment to glance at the sedate landscape far below, of patchwork meadows and trim woods painted with sun-glittering greens and yellows, and the still more distant corduroy-green, white capped ridges of wind-ruffled, sparkling sea. Then once again the aeroplane was trimmed level and pressed firmly into a dive.

Earlier, the P.V.7 had suffered from torsionally weak wings which had twisted in opposition to the ailerons, making control at speed ineffective until a special pyramid structure of bracing had been introduced. On one flight the machine had lost a port wheel and been damaged, nosing over on landing. But after the accumulation of seventy hours of tests, breaking the machine in like a pony, it seemed reasonable to dispense with further preamble and use this as the final check. I did not know that since take-off a telegram had arrived at the Aircraft and Armament Experimental Establishment, Martlesham Heath, from which I was doing the tests, saying: 'Postpone flight. Strength requirements not met at proposed air loading.'

I put the nose down more steeply still, then steadied whilst the speed built up.

Time died, as it does in all moments of concentration, leaving only the emptiness of the universe. A part of the vibrating roar of the engine and the harsh shriek of air, I hurtled downwards, a primitive, puny speck of life, dropping with frozen calculation into oblivion. Yet, at the spark of my will, at the instantaneous surge of my chemical and physical reactions, I could control my destiny: turn the madness of headlong annihilation into the sane, swift path of level flight.

I watched the airspeed rise: 180–200–210–215 m.p.h. I glanced at the revs, checked that they had not reached their limit. My gaze returned to the airspeed indicator, but in that moment the tension of the structure seemed to strain beyond its bounds. The aeroplane gave a startled, erratic dip. I snatched back the throttle.

A muffled report was followed by a thud. The nose heeled, the horizon tilted wildly and I was being swept down relentlessly, down and over and down and down, to a crescendo of shrieking slipstream.

I knew no realization of what had happened: no reasoning; no fear; only a stillness in which a great bell tolled: 'Out, out, out.'

Of their own will my hands pulled open the window against which the sideways tilt had swayed me, and a rush of air beckoned freedom. But I was trapped. Against the steep incline of the machine's attitude there was no possibility of pushing back the safety hatch above me. There was desperate need to hurry, and yet no hurry. Time was waiting, but two seconds had gone. Mechanically I tugged at the quick release pin of the seat harness and leaned across the low sill of the window. The hurricane slap of the wind dragged at me, helping me out of the living tomb. My legs banged against the fuselage and I was free.

In a blue bed of divine softness I rested on my back, and curiously watched a hundred glittering fragments desecrating the exquisite colour of the sky above. Those must be the bits of the aeroplane. The realization brought the comforting reassurance that I had not abandoned the machine too soon. From my lotus bed I heard the swan song of the main wreckage, a heavy fragment of cabin and fuselage without a tail and bearing only a single wing, hurtling to earth in a loud moan of swiftly rising pitch. Then it receded from me until I was left in quietness that was absolute.

I turned my head and saw far away England's green and pleasant land. It lay so tranquil, sun-enchanted, and so still. The poignance of its untroubled spirit reached up to me as though I bent above a sleeping child: yet I was the child,

falling to the haven of a mother's arms. I began to tumble
backwards. My legs lifted high against the sky, and my arms
were flung wide open to the world I loved. Without rancour
or apprehension, I thought that it was an exquisite day on
which to have to die, and sad that I should see its bright
loveliness no more. Of course there was still a chance. A
possibility of escape from the logical conclusion of the fall was
strapped to my back in the guise of the borrowed parachute.
But, somersaulting slowly through the air, I placed no real
trust in its ability to reprieve me. In fact I was so full of doubts
that for some seconds I put off the final action that might
irrevocably confirm my fears. At length of their own accord my
fingers groped into the canvas pocket on my chest, gripped the
metal release handle and tugged. A thin wire cable slid from
its tube and, because there was no sense of falling, inexplicably
pointed vertically upwards from my hand.

Nothing happened. It was stupid to use a borrowed parachute. The world and the sunlight looked so lovely, so
exquisite; but now there were so many things that would
never be done. A great hush surrounded me.

In the reconciled peace, the jerk when it came was all the
more violent and unexpected. The leisurely, lazy motion
ended abruptly and I was snatched upright. Bands of webbing
rose from my shoulders like magic beanstalks to the heavens.
I looked up. Making the blue still deeper by contrast was the
billowing satin mushroom of the parachute canopy swaying
on silken strings. I looked down. Cautiously I reached up to
the webbing and held the straps as though I was rocking upon
a swing. A long way below I saw the silver fragments of the
wrecked aeroplane tumbling like windblown pieces of paper.

Like floating thistledown I hung suspended above the
quilted landscape of high summer, and watched with unimpeded vision the trees and fields and threading roads. The
inimical force of death was forgotten in the tremendous
prodigality of life in cultivated land and wilderness alike.

Down there birds sang in chains of song: creatures browsed

contentedly: men and women loved. Down there in the calm, still night one opened wide the windows and gazed upwards to the stars. Yet of you and you who are so close to me, I never thought. You were all part of life, sharing its stream with the tree-top doves, the murmuring bees, the butterflies poised above bright flowers, the hare, the fox, and every other creature of earth and sky and sea. Everything was part and parcel of universal life, and life itself the symbol of a limitless and eternal power.

High in the sky I hung from my gossamer trapeze, the sun filling its silken warp and weft of threads with the white and rounded beauty of a summer cloud. Far beneath, fields slid past with swift assurance and hid in distance. Soon this quiet descent must end—but where?

With sudden comprehension I realized my back was turned to the drift, and the speed of my sailing on the arms of a gale was much too swift. It was imperative to reverse before all height was lost, or I should land as if I had stepped backwards from an express train.

Memorized as part of the drill a pilot must meticulously make, however unreal the eventuality may seem, the words of the parachute instruction book came back to me: *Pull down the shroud lines on one side to collapse the canopy and give the others a quick jerk in the direction of the intended turn.*

I pulled the left webbing. Like the frilled edge of a sea-anemone the lip of the parachute undulated, then collapsed as I increased my pull. Startled, I let go the webbing. The canopy filled. I glanced down. How much bigger the trees had become.

I must forget how quickly we are dropping and try again more methodically. Pull. Twirl with the right hand. Slowly the parachute rotates. I am a quarter round; then slowly it spins back. Surely the leaves on the trees are very clear? Don't look down again. Pull ... twirl ... twirl again ... and again ... that's better. Pull once more ... twirl the lines with a jerk ... round ... round until at last I am facing the opposite direction, sliding down the invisible slope of air, as gently as a feather falling to the ground.

Before, the earth had been widespread and beautifully remote, but now the patterns were breaking down into the intimacy of fields, sun-scorched grass, purple heaths and the swaying pines. I saw men reaping, cattle resting in the shade, a woman in a flower-filled garden—all unaware of my silent fall. A blue and white magpie went dipping and rising across a yellow field, bounded by telephone wires and railway lines—and far away trailed the smoke plume of an approaching train. Ironic if after surviving so much I was knocked unconscious on the track.

Two hundred feet up, I swept over the near hedge of a field of stubble. And now I knew how misleading was the gradual slant of descent provided by the drift of the gale. I was dropping fast. The ground rushed brutally towards me. Obeying the recollected instructions of the book, I hung limply, but the earth hit hard as if I had fallen from a rooftop on to a concrete yard. Pain seared through one leg and jolted my back, and I collapsed face downward, to be dragged a hundred yards across the bristling corn-stubble until the wind-filled canopy was held by a bramble hedge ten yards from the railway. Then all was still except for the rushing wind.

Spread-eagled and breathless, I lay unmoving, gloriously sustained by the firm, surprising upthrust of the earth. I was upheld and comforted, my spirit immeasurably assuaged. Through the stillness came the homely reassurance of many mingled sounds: the wind, faint gruff voices, a dog barking, the hum of a tractor, the rattle of a harness. . . . Such things were still mine. I had begun to live on borrowed time. Henceforth my indebtedness was the gift of days, and months, and years, like stepping stones reaching towards some different end.

CHAPTER VII

Tomorrow May Not Come

In the rich reality of the sunlit countryside through which I
motored one spring morning in 1935, a deep sense of unity
with every manifestation of existence enfolded me, and drove
away the shadowed thought that life burned frailer than a
candle flame with no assurance of tomorrow. So many times
had test-flying impressed this truth upon me that always now
the present moment seemed a flower whose perfection must
be savoured to the full. If nothing could be hoarded, then
more reason that the utmost must be spent. I had not learned
to float upon the tide.

I drove quietly, moving with the inevitability of fate towards
the new adventure of a particular path no man yet had
travelled. Behind were all my yesterdays, diminishing to far
nothingness like a coastline in the sun. Here was today; here
a vivid moment of being, of warmth and clear air, of eagerness
and rise of sap stirring the countryside and pulsing in my

veins. But tomorrow? Maybe no tomorrow—for it could be that the end of time was already but a short two hours distant.

This awareness brought no discord. I felt no more than curiosity mingling with the hope that this evening I would once more see the sunset suffuse the golden west with promise. It was not self-abnegation. I knew the hazards and was conditioned to them. If there were odds then to accept them was payment for a viewpoint of the world I would never willingly have forgone. Though life was the essence of all I wanted, to love and live were the aspiration for which every man would die, and to risk death was to know oneself and find unquenchable assurance. Even though my mind inconsequently glanced at the blank window of the future it was the living moment that enthralled and counted. I wanted to hold its transience closer—to embrace it and never let it go. Perhaps tomorrow I should have earned the right to linger: maybe tomorrow. Slowly I drove on.

Presently the black corrugated hangars of my destination rose above the curving downs, and in a few more minutes I was passing through the white painted gates of the R.A.F. aerodrome where our latest experimental aeroplane had been assembled for its first test flight. At once the dream-world of speculation was forgotten, I was too busy thinking in terms of techniques, and the routine check and counter check, buoyed with confidence in the knowledge that this machine that I was to fly represented the work of many men in different spheres of thought and practice. Each had added his impress of conviction to the design and manufacture of every part. I had followed step by step the argument and growth of this unique aeroplane, the Pterodactyl. I had listened to the aerodynamicists; watched the wind-tunnel tests of replica models with which stability and control had been assessed; studied the conclusions of the stressmen, and learnt the margins of strength. Day after day I had paused at the boards of the draughtsmen and seen the fantastic design develop in complexity from the first simple, tentative outline. Meanwhile, in the experimental shop craftsmen were transcribing the pictured shapes of thousands of parts into the three-dimensional forms of manipulated metal, that were presently assembled into other shapes and became the components of the wings to sustain me and the structure in which I would sit.

In conception this aeroplane was entirely different from the biplanes in almost universal use both as fighters and transports. Twenty years hence its boomerang, tail-less shape would still seem a little audacious and certainly rare, but in this period of its construction we already thought of it as the prototype of swift escort fighters more lethal than anything known because of the unrestricted field of fire afforded to the aft turret gun with which the short nacelle terminated.

Geoffrey Hill, the engineer who conceived the Pterodactyl, had been a pilot of the First World War, and had himself built and flown, some years earlier, the initial 20 h.p. experimental spruce and balsa-wood version of this flying wing. Two successive developments of the machine had been investigated on many research flights by my colleagues and myself, and with one, the first loops, rolls and spins of any tail-less type in the world had been made. So, with some years' experience behind the design, it was felt that this time a fully operational weapon could be achieved incorporating advanced ideas in the technique of the metal construction that was generally replacing wood. Powered by the latest Rolls-Royce steam-cooled engine, this unconventional aeroplane should prove comparable in performance and ability with the newest fighters built to the standard formula of shape.

Even the mathematics of stressing swept-back and twisted wings had to be evolved and proved for this new creation. Every aspect of the design, from wing-tip controllers to the bicycle-type undercarriage, presented a new problem to be solved by the logic of calculation and endorsed with practical tests. Yet because new ground was being broken no man could be quite sure. If the fundamental assumptions were incorrect the whole fabric of calculation could tumble to the ground. So, indeed, the first taxying trials had demonstrated.

Earlier in the year the Pterodactyl had been assembled at the factory. Silver-bright with newness she had crouched on the turf, the sweeping wings drooping backwards like a rabbit's ears. No one then suspected that similar swept-back wings would prove the means of attaining the undreamed-of speed of sound.

Below the centre-section of the long top main wing hung the grey nacelle, with wooden propeller and steam-cooled engine in its nose. At the bottom of this short, metal-covered body a small, dagger-like wing projected on each side. From the tips rose streamlined struts, in the form of a Vee, holding down the big wing from the middle of its span. Behind the lower wing tips, long, metal skids slanted to the ground to prop the aeroplane laterally.

On the day that the preliminaries of engine test and final inspection ended, I had climbed into the open cockpit. The sight of swept-back wings on either side was strange only because a view of green turf replaced the walls of the shed in which I had spent many hours familiarizing myself with the arrangements of the growing aeroplane. Eighteen months earlier I myself had planned the cockpit lay-out in the full-scale wooden mock-up.

By the time any new machine is ready for its first flight the pilot has become so used to the layout, view and shape, that they are almost a part of the furniture of his daily existence, and far less harassing than an aeroplane which may have been long in production but has not been flown by him before.

So when I looked through the narrow triplex windscreen, along the slanting nose past the grey painted propeller blades, the view was all that I had expected, and better than that of any previous fighter. A glance behind confirmed the unencumbered field of fire which the tail-less design afforded.

Mechanics standing on the stub wings wound the engine

cranks, jerking round the wide propeller blades. With a puff of smoke the engine breathed life into the structure of the aeroplane, and the propeller spun into a haze. My own checks completed, I waved away the single front wheel chock, opened the throttle a little, and carefully taxied off. A crowd of workmen and designers awaited in silence the metamorphosis of their patient creation from the stillness of a static structure to an animation that was more than mechanism.

Up the gentle slope of the home aerodrome, over the familiar brow and on to the level top, the aeroplane rumbled, its twin-skids grating and the front wheel kicking but responding to the press of my foot on the rudder bar. I pointed the flying wing down the length of the bumpy field, opened the engine a little more and accelerated into a tentative run to get the feel of the machine.

After 400 yards I slowed, and paused, and turned. Once again the engine was mildly speeded. The aeroplane travelled a bare three yards before a sharp metallic scrunch, and a drunken, sideways lurch made me instinctively flip off the switch. Twisted and broken, the port wing crashed to the ground.

The crowd came running. I gave a mental sigh. This time the experiment of flight might prove more difficult than usual.

But that was months ago. The mishap had been investigated and the fault located. An unexpected eccentricity in the manner in which a load acted on a secondary strut had set off a train of failures like a falling house of cards. Since then test and re-test with laboratory rigs had been made to simulate the theoretical loads and determine the ultimate strength. No doubt lingered, and the wing had been rebuilt. As far as human integrity could make her, the Pterodactyl was structurally sound. Now, with sunlight burnishing the polished aluminium and silver paint, her angled wings hunched in the softly moving air, she was ready for me to discover whether her flight was perfect, too.

Fitters, designers, and inspectors waited with me as I

checked the loading. A faint threat of tension and restraint
made talk brittle. At last they gave me the signal. I fastened
on the newly packed parachute, took my helmet and goggles,
and climbed into the tight cockpit. The sea of faces became
forgotten as I strapped myself to the seat harness. Time and
date were written on my knee pad. Instruments and controls
were checked and set. We were ready.

The galvanic force of a quite different and expectant vitality
began to flow, turning thought and reaction to knife-edge
intensity. I nodded to the men to crank the engine. A moment
later the aeroplane surged into life which fused and blended
with my own.

Chocks away, I taxied cautiously towards the leeward corner
of the aerodrome, conscious of the tight-rope walking motion
of the tandem wheels. Already the sense of the machine's
inertia and the manner of its response to the throttle seemed
characteristics with which I had long been familiar. Wing-tips
trailing backward, wing-skids rumbling, propeller glittering
in the sun, the flying wing moved over the aerodrome, drawing
a tendril of exhaust across the grass.

I turned by the hedge, steadily opened the engine, and let
the acceleration build up in a half-mile run along the turf.
We paused, then slowly returned.

In the background of my senses there was faint mockery
at the bold front of the moment. Though I might open the
throttle with deliberation, this cynical inner voice jeered that
this was no more than a prelude, bereft of any intention of
making a full flight.

I made two more taxying runs at increasing speed, feeling
the trim and stability, and trying to gauge the balance and
effectiveness of the controls. Groping in the unknown I could
only surmise that nothing was obviously wrong. A still faster
run was made, and this time the machine gathered flying speed
and lifted into the air, allowing me a brief feel of the response
of the controls before I throttled back and the machine sank
to the ground. After another hop I was satisfied that all was

well and taxied in for the cowling to be removed and a last check made of structure and controls.

Half an hour later I taxied out again. At last the moment had arrived: the inevitable outcome of a long sequence of actions. I was buoyed with pride at being entrusted with a machine that represented the pinnacle of a new achievement, and exaltation at the privilege of my own function in the exploration of new worlds. Yet as I looked up from a last scrutiny of the instruments and gazed across the space of turf stretching ahead, I was pricked with the uneasy recollection that in its first trials this fantastic aeroplane had crumpled up. Instantly I tried to smother doubt with belief in the work of the very people whose logic had been proved fallible. Ah! whispered the voice, but how trivial a mistake! How great the consequence, came the echo.

I, of whom the arguing voices were a part, felt nevertheless remote from the debate. My life's force was preparing for the uplifting prospect of the next moment's movement into flight. I opened the throttle. Power poured through the machine, mounting in the quivering structure, mounting in my heart. The flying wing began to move.

Hold the control column lightly and still. Steer straight. Let her feel her way into the air. She is swinging: check quickly with the rudder: more—a bit sluggish in responding. Steering is difficult on this bumpy ground. A glance to left and right shows the wings equally level, their skids trailing through the grass tops. Keep straight, straight—full rudder; centre; keep her straight; ignore the curious lurching. The grating of the skids has vanished. From the corner of my eyes I see they have lifted. The sweeping wings are already leaning on the air and letting it carry their weight.

Through the gale of wind that batters at my goggles, I watch the boundary hedge drawing swiftly nearer. This is the datum line of my consciousness. The turf is blurred and racing. Suddenly it is too late to stop. I can feel the animation of the wings, and the sleepy strength of the controls—but she *must* take-off, or a few seconds more will find her crashing through the hedge, the labour of thousand upon thousand of hours broken in an instant to crumpled wreckage: her story finished before it has begun.

With an action that I knew, for all my concern, must be inevitable, a last jolt bid farewell from the ground and the wings were lifting her free, the hedge dropping and vanishing.

An immature eagle is trying out her wings. I concentrate mind and muscle on keeping her steady, waiting with held breath whilst she climbs on a long, flat trajectory towards the safety of the higher skies.

So still were my wings in the calm air that they might have been a painted picture; yet for all I knew they could be balanced upon a knife edge, since stability depended on a twist, built into those back-tapered tips and replacing the leverage of the usual tail-plane. If the wings were too flexible and altered their incidence when a varying load was imposed anything might happen.

Very cautiously I rock the lateral control, then hastily centralize. It feels queer—as though a massive pendulum swings from a frail lever; so what will it do when I try more vigorously or move the combined ailerons and elevators together in a compound change of flight path? Presently is soon enough for the answer—presently, when there is ample height to give time for a parachute descent; meanwhile, keep her level and straight.

I took a breath. The noise of the engine thundered into my consciousness, silencing all thought for a few moments before it dwindled once more to an insulating drone that shielded me from sense and reality. With the untried fearlessness of all new and inexperienced things the aeroplane climbed onwards, as though it had long been familiar with the engulfment of air that streamed and eddied over the taut fabric of the silver wings. My senses tuned to high pitch I watched for the slightest movement, wondering with queer dispassion how the machine would react to the control forces that I must soon impose. With tolerant curiosity I looked towards the things that unhurrying time would bring for good or ill.

The green drift of fields and trees beneath the diminutive lower wing gave place to downlands unfolding and expanding as the aeroplane climbed still higher. There was no recollection that I had driven across them earlier in the day. In this springtime hour the land rested as though enchanted, serene with knowledge of the past and waiting with long patience for the years to come.

The altimeter moved its slow hand towards the 4,000 feet calibration at which the controls would be cautiously moved to gain level flight, and the critical moments would begin.

Without further thought we were there, and the aeroplane was level. Speed increased steadily whilst I progressively sensed its safety through the nerves of the flying controls. I removed my hand from the column. The flying wing flew level unfalteringly for many seconds. There was nothing immediately dangerous, and stability was good enough for further tests. I eased a little in my straps and breathed more freely.

I throttled back and slowed down; then scaled upward through the range of speeds again, this time rocking her laterally so that the wings drew tilted lines across the horizon at steeper and steeper inclination. The reaction was a little cumbersome, as though she was big and heavy in this plane, compared with the quick responses when I rocked her fore and aft. It was good enough to warrant further exploration.

I studied every visible section of the structure. All was well. Once more I slowed the aeroplane, only to open the engine again and try the rudder through the range of speeds. She answered a little drunkenly, rolling as she did so. Something not right here. Leave it to check another day.

One further, vital step into the unknown remained. Before a landing could be made with the precision of confident knowledge, I had to know what would happen when the airflow over the wings began to break down at the machine's slowest flying speed. It was here that measurements on models tested in the wind tunnel had indicated possible danger. Slats like movable pinion feathers had been fitted at the edge of each wing-tip to retain the lift and keep each side in balance. As the sweep-back of the wings placed the slats towards the rear of the machine, the leverage of the displaced lift was expected to tilt the flying wing downwards. The model tests, however, had shown that because of its forward position the centre section might retain excess lift and overcome this stabilizing force. In this case the aeroplane would tilt uncontrollably upward to the stall, and fall into an irrecoverable spin. My mind concentrated into cold calculation and I began to close the throttle. The silent seconds dragged into an empty pause.

I raised the controller flaps higher and higher but there was little marked change in the incidence of the wing. Only the descending needle of the airspeed indicator showed that the attitude was altering steeply in relation to the invisible path of air. The balance of the machine grew precarious. The controls were ominously devoid of feel. Speed wavered at 45 m.p.h. I waited: waited for danger to leap, for violence, for the uncontrollable unexpected.

The control column jerked backward in my hand: the wing-tip controllers flicked fully up, overbalancing and locking in position.

Now what? Spin? Stability intolerable? Controls impossible to return to normal? Wait . . . the A.S.I. stayed steady at 45. The altimeter was unwinding like a clock. I took a deep breath of relief. The machine was fully stalled, dropping with the measured safety of a parachute.

I pushed lightly on the column. A little stickily the big controllers moved down, and down, too, went the nose. I opened up the engine and, descending under easy control, the flying wing swiftly gathered speed. There was new safety in my wings—and the countryside smiled up, beckoning me home.

Enough for a first flight. Land now, and have the aeroplane examined before doing any more. Sufficient for the moment that she flew controllably and more or less conventionally. Next time she must be dived, but not today; not today for any of the unsuspected perils that I later discovered in the flexible ends of the tapered wings, which faster speeds were to twist upwards so that the downward extent of the controls was neutralized.

Back to the aerodrome I headed the machine. The triangular green space of the landing ground made it easy to find among the hedgerow-bordered rectangles of many small fields. With sweeping curves from left to right, the aeroplane dropped towards it, the canted tail-less wings glinting with sunlight against the background sky. Many a man must be staring up amazed at this bat-like creature sliding through the heavens. Yet only when I turned in my seat and looked straight down the empty abyss behind, was it possible to realize that my aeroplane was in any way abnormal. I was conditioned to the strangeness.

Like a familiar, the Pterodactyl responded to my slightest wish transmitted by the lightest touch of foot and hand. Boldly it swept round the aerodrome, where, in the shadow of the hangars, stood the little group I had left twenty minutes earlier. To show that all was well I dived towards them, sweeping past close enough to recognize a few faces; then up in a wide climbing turn, gauging the one right moment which on closing the throttle would bring the trajectory of my glide within just sufficient clearance of the hedge to let the aeroplane settle on the grass close to those men who had dreamed of her, and built her, and at last had watched her fly.

Through the quiet evening I drove slowly home. It was good to linger. This was my country, my heritage of beauty, my own good earth. Here was the sentient world where my life was rooted. This was the world that held me in willing bondage, and this moment of renewal made me still more its own. Yet it was not reaction from any sense of strain that bonded me thus closer. The quiet curiosity and conjecture of what may happen in the unknown of any flying venture has no apprehension. Instead, one's insignificance in the vast amphitheatre of sky seems to draw a power beyond mortality, which restores the vision of the earth to greater brilliance.

CHAPTER VIII

Soaring Sea Quest

On a day of boisterous wind in 1936, I took my newly completed sailplane Pegasus to the Dorset hills for her first flight. We assembled her quickly on a grassy path cut through heather and tall bracken. Her slender, translucent wings swayed to the wind off the sea as we tied her to the ground.

On other occasions when I had been the first to take an untried machine into the skies it had been different. With them it was as though time held me in suspense, freezing my mind to stillness. Speculation was subordinated while I waited with a sense almost of leisure and release for the moment with which my destiny was inextricably bound. But not this time. Instead this was adventure and attainment. These were my wings, grown from my imagining, made with my hands, in emulation of the gulls that I had watched soaring above the cliff edge.

Once again I had discovered how difficult it was to translate dreams into reality. Compared with the aeroplanes that I flew, and at one time helped to design, the project of the sailplane had seemed a simple undertaking. Months had grown into years, however, since the first lines of manlifting wings, long and tapering like a gull's, appeared on my drawing board.

Slide-rule and sheets of figures, curves of lift and drag, tabled weights and moments, calculated stresses of wood and steel, presently confirmed that the miniature sailplane of my sketch would have adequate performance and strength. With a span of thirty-four feet and a mere ninety square feet of wing surface, my Pegasus was the smallest sailplane in the world. It had to be small because it was built in my bedroom. It had to be small for cheapness and easy transport. Because it was so small I would feel at one with this £30 fantasy of varnished spruce, thin plywood and doped linen.

I wriggled into the cramped cockpit, to which the leading edge of the wing acted as a roof. I strapped myself in, and felt for the miniature rudder-bar.

I tested the controls gently, feeling the response of their surfaces through the lightly rasping wires and levers. The elastic launching ropes were stretched, the machine poised for flight. Everything was very still, except the mounting wind, and in that instant, while everybody waited, my own identity merged with the mechanism of the sailplane.

'Release!' I shouted at last.

A stir of life throbbed through the wooden structure—or was it my own heart beating? For a moment Pegasus clung to the earth, and then surged forward, parting the air like an arrow, and lifted serenely on the breeze. The launching rope fell from the nose—and she was free, leaping the grey stone wall that barred the pathway's end, revealing the hill's steep edge that bordered the sea.

Hovering kestrel-like, rocking and swaying, the sailplane soared higher on the wind, until the launching field was the size of a fallen leaf, and the great crescent hill of Kimmeridge dwarfed to a wide arched brow edging a tableland, behind which rose the Purbeck Hills. Turbulence flexed the wings in jerks and swept the machine up and down with sudden surges, whilst my feet and hands worked endlessly at the controls with the same instinct as a gull twisting and flexing pinions and tail to hold a level path.

I swayed the glider into the wind and began to follow the curving course of the hill to the west. As Pegasus continued to rise the view spread further until I could see, on one side, the brown moors surrounding the glittering waters of Poole and, on the other, the crouching form of Portland, blue and hazed, dwarfed by the great cliffs towards which I steered.

Air flowed round the cockpit with the swish of waves on a sandy shore. Although the minute size—the wing chord ranged from 1 foot 6 inches to a maximum of 3 feet 3 inches—gave the sailplane a very small inertia and made the control column and rudder bar unusually sensitive, I instantly became accustomed to the handling. Glider, fighter, bomber—the flying was the same, but the absence of mechanical power made the floating without visible reason seem a miracle. The lack of vibration, too, gave a sensation of natural bird-like flight that I was not to discover in an aeroplane until the advent of the jet engine.

Foot by foot the sailplane crept forward, steadily climbing. Presently it was over a thousand feet high. Directly beneath the polished, narrow, mahogany box in which I sat, the hill curved rapidly towards the sea, and it became easy to appreciate the steep rise and fall of the distant cliffs. Like a series of monsters they ranged westward, humped high in the sky. They were my stepping-stones, where the oblique wind would be deflected upward and let me rise enough to jump the gaps between.

But even while assessing the chance of reaching Flowersbarrow, and leaping the Bindon, the sailplane began to drop. We had reached the point where the curving hill was dead across the wind, and we were flying in the downdraught of the lee. Within a minute half our height had gone. I threw a desperate glance towards Brandy Bay where great cliffs faced the wind, and promised safety. Could we reach it? The sailplane was dropping like a stone. I pushed the nose down to gain more speed against the strong headwind, but the sailplane merely dived at the hill and made little progress. Each tuft of grass grew clear, becoming a signal of danger.

The tussle ended almost as soon as it had begun. One moment Pegasus was scraping the hill; the next it had reached the cliffs and swept in front of the wall of Gad, well below the top, flying a bare 200 feet above the sea. Across the water gulls by the hundred rose in a cloud from every niche and pinnacle at the sudden silent arrival of the sailplane. Where they flew, could not my sailplane fly also? I turned closer in to the cliff, and as though lifted by a giant hand, the machine surged upward again.

Rising unsteadily on the turbulent wind, I edged Pegasus along the chasm of the cliff. The gulls scattered before my path, and a raven sped on hunched wings towards Worbarrow Bay. With undeviating flight he reached the far cliffs, and I grew assured.

There was time to examine microscopically the scarred and primitive cliff whilst we climbed past its face until the sailplane lifted above the grassy top and gained safety with a thousand feet of height. The panorama of cliff and hill spread out distantly below me again, and I was no longer simply a mortal crouched in a flimsy machine, but a bird in harmony with sky and water, rocks and turf-grown chalk.

Gad slid from sight, and Worbarrow Bay curved around the white-maned sea, with Tyneham's green valley peacefully beneath my wing. High above the sea rose Flowerbarrow of the ancient dead, crowning the shorn tip of the Purbeck Hills at the coastline of today. From its white side a hurricane of air lifted me a further 500 feet in one tempestuous leap, and sent me easily across the long gap of Arish Mell. Slowly we floated high above a deep valley crowned by the broad sweep of hills fashioned from the age old conflict of land and sea. Long generations of man had lived their span upon that turf leaving no more relic than the cultivated shaping of their land and the still more ancient symbols of grass-grown funeral mounds.

Scarlet and russet in the sun, Rings Hill drew abreast. From the magic of Flowerbarrow I looked far across the brilliant heaths of Stowborough to blue distances rising fold on fold. There, brothers of ancient men who made the citadel below long held domain in mightier castles carved in dazzling chalk. My wings swept lightly across 2,000 years, and brushed 2,000 more—and all the while I gazed at a world sculptured in flowing modulations of form and colour, austerely lovely, drowsy with infinite peace.

Abruptly, the tranquil moment fled. A succession of eddying bumps jerked me into reality. Pegasus had flown into the downdraught over the nearby edge of Bindon ridge. The variometer signalled a red warning of descent. My only chance was to steer seaward towards Mupe Rocks, where the wind struck the coast at an angle and sprayed into upward currents. It was a race between the machine's ability to move forward, and her quick gliding descent. On either side spread the white streaked sea with Mupe Rocks small and grey in the distance. The waves grew larger and success or failure hung in the balance. Then the shadow of my wings touched the long, south face of Bindon.

Air swept past at nearly forty miles an hour as Pegasus continued to drop. We were thirty feet from the side of the hill before I felt the first faint up-current. It was too weak to sustain us. Closer and closer I edged until one pointed wing tip was a bare five feet from gable sloped turf. Beneath the other wing lay a great void. But we had won. Inch by inch Pegasus rose to surmount the long ridge. My eyes drew level with the top. A minute later I could have skimmed across to make a landing below the other side. Instead we sailed on and on, rocking with little jerks, rising and falling, but always making good our height.

Lulworth Cove, blue and unruffled in the shelters of the tall limestone cliffs, came into sight. I could see the steep cliffs of Hanbury Tout. An easy jump: and once there, the whole coast offered a buoyant path of up-currents—White Nothe where Llewelyn Powys used to walk; White Horse Hill near Georgian Weymouth; White Sheet above grey Portisham; the level Burton Cliffs; Golden Cap, the highest cliff in the south—and on and on until the land curved so far seaward that it would no longer hold the wind.

Behind me the westward end of Bindon stood like a chalk precipice, against which the full blast of the wind was forced upward in a mighty current, on which the gleaming wings of a host of gulls circled a thousand feet above me.

I looked at my own wooden wings and felt pride in them. Little more than matchsticks, they were strong enough to give victory against the forces of the atmosphere. Lightly I rocked the control column, appreciating the instant and delicate response. There was rapture in the assured movement of Pegasus and the power of her flight tricked me anew into a sense of affinity with the wild scene of sea and sky and land.

Far below, groups of people stood, watching us. I looked down in triumph at the scores of upturned faces. I was a god; the world a plaything. The sea-wind stung my face and pressed against my wings as though I were indeed a bird. Back and forth Pegasus shuttled across the cove until presently we poised motionless above the coastguard station on its hill.

Cliffs drifted by. One moment Hanbury Tout was far below; the next, we were sweeping seaward of the arch of Durdle Door. So slowly that at times the machine seemed to be making no progress at all, we travelled the plunging line of cliffs.

Waves broke into flying foam upon the pebbles. Gulls soared in drifting groups upon the wind. I rose higher and higher above the white and shining coast.

At White Nothe, where the weathered chalk of the coast changed to lower walls of sand, I met the strongest current of the journey from the wind falling unimpeded upon the westward flank and rushing skywards with an even more furious impetus than that of the updraught at Bindon. Pegasus rose higher and higher until a thin mist unexpectedly dimmed the view; and then suddenly we were enveloped in thick white cloud.

In the small cramped cockpit the cloud was more claustrophobic than I had ever known. I had no compass for direction, no gyro-controlled artificial horizon to give balance, no turn indicator, not even a spirit cross-level—only an altimeter and airspeed indicator, and the draught on my face and my leaping senses, to guide an even keel. I sat, tense and breathless, trying not to move the controls.

Even to a low-powered aeroplane that sixty-feet depth of cloud would be no more than a single second's mist—and merely a steamy discoloration scarcely observed in the passage of a sonic fighter—but to me the long seconds of slow climb were filled with imminence of danger and the emptiness of suspense. The world of love and beauty, of green life and sunlight, the world that was my gift of time, was obliterated as though it had never been. There was only the drab, suffocating whiteness to fill the waiting.

As though sprung from my longing, a soft radiance appeared. The light grew firm, the cloud thinned. My senses re-orientated themselves. Tugging free of the last silvered vapours Pegasus burst into the startling brilliance of unimpeded sky. Higher still we rose until the cloud, that had been vast enough to imprison me, was a thin, white floor so narrow that it hid only the tall cliffs of the Nothe and the immediate vicinity of sea.

The needle of the altimeter slowed and stopped at 2,700 feet. The green pip of the variometer dropped to zero, and the red indicated a steady descent of a foot a second. With time in plenty, and height in hand, I scanned the panorama of coast and hill assessing the best course. A few scattered clouds promised thermal lift, but I feared their dank imprisonment, and looked for a route to my left where cliffs stretched gently towards Portland Bill. The white wings of gulls sailing along the shore promised well.

By following that route, where the wind blew against the cliffs, I should be able to jump the low isthmus of Portland, and reach a better line of windward hills, backing the Chesil beach, and offering an easy flight to Devon.

A mile passed in comparative safety, but we were dropping gently and in the next mile lost 500 feet. The descent seemed logical enough, the cliffs lay lower, so the ceiling of the up-currents must be less. At first I had no qualms as we dropped slowly in our struggle towards the distant blue of Weymouth.

But soon Pegasus was sinking faster, at the rate of five feet a second. The altimeter dropped to 1,500 feet and then to 1,000. I considered turning back and racing downwind to the safety of White Nothe. It was within three minutes' reach, but in the urgency of the moment that seemed too far away. I held my course.

Hardly was my mind made up before I realized I had erred —and then it was too late. The bumpy cliff top was so close it almost touched my skid. A weak up-current held me for an uneasy mile, and I comforted myself with the thought that if the cliff wind failed, I should have a few seconds in which to turn for the fields on top. Yet the contingency felt too unreal to be disturbing, although the wind under my tail would mean touching down at 60 m.p.h. and a certain crash in the confined space available.

A gully gave a brief glimpse of an inn set near the beach, and a road winding through a rough valley where a field might offer safe landing—then it had gone and the cliffs held my attention. Pegasus was uncomfortably low.

A mile ahead taller cliffs promised safety, but the variometer was showing descent at two feet a second. In one minute more my sinking path would join the ground. With instinct stronger than thought, I banked steeply round, ruddering in a tight circle across the sea, turning back for the hidden gap of Osmington half a mile behind. Our speed increased with the following wind and I searched anxiously for the narrow gap.

In thirty racing seconds we reached the tiny gully, and I banked steeply into the gap, the cliffs looming high above. Gusts buffeted me from both sides. I had to fight to keep control until the valley opened.

On an easy curve the sailplane canted into the wind over a rough, sloping field beset with grass tussocks and brambles.

It was too late to change. An errant gust shook the wings as I levelled off, skimming a foot above the field. The skid bumped on a molehill, hidden in the grass, and Pegasus rose a little, flying free ten paces, before touching down again, slithering roughly to a stop and settling on one wing.

I took off my helmet, and sat unmoving in the cramped cockpit. In this sheltered place the wind sang a gentle note. A grasshopper chirped. Subdued on the breeze came the dull roar and thud of distant breakers.

It had taken us ninety minutes to fly twelve miles of coast, but I was already forgetful of this conquest that I had planned so long to make. My heart was filled with a vague consciousness that somewhere I had nearly seen the lifted veil: that I, who adored and desired, was an integral part of the natural universe, like the wind and the tide and the warmth of the sun.

CHAPTER IX

End of the Beginning

In soft sunshine I walked across the heather clumps fringing the tarmac apron at the Aircraft and Armament Experimental Establishment at Martlesham to gain experience of the latest conception in fighters. This was the shape of things to come, the low wing cantilever monoplane, made possible by retractable undercarriages and new alloys for the fashioning of smooth-skinned structures combining lightness and strength.

For long years scientists and engineers had dreamed of this approach to the ideal bird form; and now, in the year 1938, the urgency of the international situation was removing the last prejudice in favour of the traditional biplane, and speeding the evolution of the most famous fighter of the Second World War.

The temporary, fixed-pitch, wooden propeller did nothing

to mar the ultra-modern appearance of the lineal descendant of the S6B, Schneider trophy winner. Smooth as ice and glistening in blue-grey paint, the lines of the Spitfire looked exactly right, a testimony to the awe-inspiring repute of its speed.

The interest and the hidden questioning at flying an unknown machine steeled into the fatalistic confidence all pilots recognize, as I clambered on to the wing and squeezed gingerly down between the semi-circular windscreen and the pushed-back hood. Settling into the tight-fitting metal shell, I let the hard newness of its touch filter through my consciousness. There was strangeness in the upright sitting posture, in the queer smell of different paint and the sharp tang of hydraulic fluid. My glance outside took in the relationship of wing and ground. The view ahead was blocked by the broad, flat nose.

Everything felt disturbingly unfamiliar, but the sooner I made a start, the sooner that would pass.

A puff of thick blue smoke and the engine ticked over with a rumbling rattle. Power tremendously greater than usual lay at the touch of the throttle. The rest was a jumble of unaccustomed impressions: the rolling gait of the narrow undercarriage as I taxied out; the dropping wing and emphatic swing as the over-coarse fixed-pitch propeller laboriously gripped the air, dragging the machine into a run faster and longer than anything I had experienced before.

Trees on the far side of the aerodrome rose like a barrier with the aeroplane held on the ground by a wing-loading nearly twice as great as that of any previous fighter. It seemed impossible that we could lift in time—and yet it was too late to stop. The echoes of my thoughts recognized a situation encountered many times before. I moved back the control. The jolting ceased, and we were airborne with nose and wing blotting out the trees, unpleasantly close though they must be. Independent of my will, the Spitfire sailed into the sky.

For some minutes I held the controls quite still, before I

groped against the heavy slipstream and closed the hood. The roar of the airflow receded, but within the confined cabin the racket of the engine grew threefold. Great waves of sound pounded my senses, as we climbed in a long, flat trajectory at an incredible speed compared with that of the Gladiator, the fastest biplane fighter of the day.

I tried to translate the meaning of the glowing green signal on the dashboard. Ah, the undercarriage! I found the actuating lever and pumped vigorously. The flight path oscillated violently with each stroke, for I could not stop my left hand on the control column from moving in sympathy. The longitudinal response seemed dangerously sensitive. For a moment I was apprehensive, but the thuds of the undercarriage legs locking into the wings reassured me with the explanation, and the machine settled into a steady flight path. Already the Spitfire's formidable personality was becoming part of my consciousness, and the slipstream hurrying past the flimsy hood epitomized the swift current of my thoughts. The thunder and din ceased to be recognizable as noise, and with an ever-recurrent trick of imagination turned into walls imprisoning me more strongly than the confines of the close-fitting hood. I looked through the perspex at the broad fields, golden again with harvest, a background stage-drop, set for a solitary actor.

Suddenly a Gladiator appeared a thousand feet above, its fixed cantilever undercarriage extended towards me like an eagle's claws, and offering the opportunity of a mock dog-fight.

I drew the stick back in the manner to which I had been long accustomed, unprepared for the lightness of control.

A vice clamped my temples, my face muscles sagged, and all was blackness. My pull on the stick must have relaxed instantly, but even so, my returning vision found the Spitfire almost vertical and the Gladiator fully two thousand feet below.

More than a thousand flights since then with many versions of the Spitfire, which we manufactured under licence at Westland, have endeared it above all others of similar vintage.

Together we ventured into every kind of weather: clawing through dark clouds and bursting from pinnacle tops into dazzling sunlight; groping with throttled engine and dead radio above shadowed ground made featureless by breathless mist; flying in the suspense of the dark spears of tempestuous rain attacking the windscreen and obliterating all view. But there were calm days, too, of summer sun, and tall clouds drifting like great ships with glowing sails; days when the skies yearned with remembrance of peaceful years and forgot they had become the medium of war.

The sense of unfettered power made it almost impossible to fly this aeroplane without sweeping into the curves of aerobatic flight: the skies whirling under wing, the earth appearing in its stead and rolling over sideways, further and further, until once again sky and earth were in their appointed place. Power surged through the metal structure as the throttle was opened fully and the fighter lifted from its dive into a racing climb to a peak that displayed five counties in the same small panel of windscreen which a few seconds earlier had framed only fields. Into the far infinity of the skies I gazed. Was the enduring permanence of the earth-world designed merely to shelter the fleeting impermanence of men? What was the answer to the riddle of this universe in which man vainly and illogically dreamed himself king?

Soon to that same earth I hastened to return, the glory of the universe discarded for the sights and sounds with which life seeks to hide its loneliness and apprehension.

Time and again, while the months mounted into years, the pattern of such flights was repeated with other types of fighters, both single-engined and twin. Always there was the closed-in solitude, thunderous noise; the rhythmic undercurrent of vibration, whirling propeller discs; the far-away patchwork of countryside indefinitely seen through cabin windows, or the obliteration of cloud thicker and more suffocating than fog.

Presently the heights would be attained: the symbol of

mankind's endeavour, a wing frozen against the endless emptiness of the metallic blue and the white blaze of the sun.

In the illusory motionlessness of high flight I would poise, crouching and still, my own heartbeats fused with the aeroplane's vibrations, and the only tangibility the sun-filled cabin, its instruments covered with light hoar frost. In that tumultuous noiselessness I was startled to hear suddenly in the earphones a voice: 'Hello Owlbird, this is Dogster. Please transmit for a fix.'

Each flight of this epoch now was worked into the pattern of a comprehensive system of ground control and radio aids; yet there were occasions when the plotting failed to differentiate between a hostile aeroplane and an allied one, and times when some airborne operation imposed radio silence.

On one such occasion early in the war I took on test a new type twin-engined fighter, the Whirlwind. Unbroken overcast at 10,000 feet sealed an all too limited sandwich of space between ground and sky for so fast a machine. Soon all that was possible at low altitude had been completed, so I radioed we were climbing to gain height above the strata for a dive. A brief, tenuous light, like smoke at the end of a tunnel, heralded our break through the clouds into the radiant blue above. I looked around. . . . Barely half a mile away was another aeroplane flying an oblique intercepting course almost at my level. I banked away to avoid collision and levelling, glanced again at the intruder. With quick shock I saw black crosses. A Hun! There was no more than irony in the four cannon projecting aggressively from my own fighter: they held no ammunition because test pilots ranked as civilians and we would have broken the rules of war.

I put the nose down plunging for the safety of cloud. Simultaneously the Me.109, impressed by guns and twin engines, jerked into a steep turn away from me and dived for the same cloud cover. We entered it at the same moment, but in opposite directions—and there I stayed a good ten minutes, for the first time in my life finding relief at flying blind.

When eventually the shadowed earth came to view it was dull and tired, as though bereft of hope at the futility of warring man. Yet as I looked around that subdued landscape, searching for the landmarks that would bring me home, I noted the first faint green of opening buds—and, dropping lower, sign upon sign of awakening spring. In my desire to escape, conviction insisted again and again that all that mattered was the safety of the earth, the essential earth in which is rooted every human aspiration; the earth that lets time spin by and discards the love of lovers, yet is always waiting to entrap newcomers to the world of love; the earth where I longed to live for today—tomorrow might not come.

CHAPTER X

Dogs of War

War brought flights in many aeroplanes besides our own fighters, that ranged across the designing genius of a half-decade marking the end of the beginning and at the same time a prelude to a new air age.

A gusty autumn wind was hurrying dull clouds over the sky as I squelched across muddy grass to where a captured Me.109 stood a little apart from some Hurricanes. Not even the orange, blue, white and red roundels with which it had been painted could disguise the fact that this dangerous-looking fighter had the typical lines of a Hun. In striving to achieve maximum performance Professor Willy Messerschmitt had made the smallest feasible aeroplane to carry the biggest possible engine—and at this early stage of the war the result looked as though it might be tricky to handle.

It was a solid little machine. Despite its massive weight, the wing had but three-quarters the area of the light aeroplanes so popular before the war. Consequently the narrow, stilt-like undercarriage gave a top-heavy appearance and the heavy wings remained unrocked by each passing squall. Some of its features were clumsy. The folding hood was one. When I climbed on to the wing-root to get into the cockpit I needed help to steady the unwieldy canopy. Nor could I shut it myself for the wind tore it from my grasp. The neat sliding hood of British fighters was a far better arrangement, enabling the pilot to open it himself and obtain a better view for taxying and take-off.

The cockpit was cramped. The windscreen was like a tunnel, with a flat front too far away, large sidepieces attached by broad frames, and the top boxed in with a curving roof. At the rear of the seat an armour-plated bulkhead rose to the top of the canopy, and bent forward to protect the head against shots from above.

The German-worded instruction plates on the dials and controls added to the strangeness. If the absence of blind-flying instruments was accepted, the cockpit had an excellent lay-out. Starting the engine brought a heave from the aeroplane as though awakening from dreams of German skies and German youths. Then it began to vibrate in phase with the motor's dull rattle.

The waiting mechanics swung their arms for warmth in the bitter air. Inside the cockpit it was cosy and secure enough to make the actions of the outside world unreal. The metamorphosis was always the same.

I waved away the chocks and started to taxi. The view was appalling, and the machine difficult to turn at the take-off point. A green light flashed from the control tower: I had no radio.

(Unlike Allied aeroplanes with normal hesitant carburettors the pick-up of this direct-injection engine was instantaneous and perfect. Up came the tail, flying cleanly when the stick was held forward. Acceleration was swift, though the undercarriage imposed a jerky lateral rocking from wheels scoring through the mud onto the hard, rough surface beneath until she rose into the air in a smooth take-off.)

I retracted the undercarriage and re-set flaps. A misty

chequer-board of ploughed fields and black hedges rotated slowly as the fighter banked towards a lighter patch in the overcast cloud. Engine noise greater than the racket of a tube train reverberated through the cockpit and vibrated the thin metal structure. But the aeroplane flew confidently. I could feel the certainty of its response even with the scarcely perceptible control movements of steady flight.

We broke through a large gap in the ragged overcast into the clearer air above. Broken cloud like muddy clumps of snow, floated upon a dingy-pearly mist.

Flying so fast a machine in such poor visibility with no radio aids was a bit worrying. I opened the clear-vision panel in the windscreen. The noise of the engine doubled—but was quickly forgotten as I began to explore the fighter's handling characteristics, in the manœuvres that were my standard programme of comparison from one machine to another. I was the human instrument against which they were calibrated: a medium of swift responses and reactions, of fears and reliefs. There were degrees, times and forces to which I could respond with unthinking ease. If these were exceeded then the aeroplane was difficult. Soon I knew that the Messerschmitt had many minor snags of control and stability, which in certain circumstances would reduce its potential as a fighter. Indeed the odds were that if the pilot could be tempted into a half-roll and dive at moderately low height he would not be able to pull out. At lower speeds or in very tight turns and violent manœuvres, the wing-tip slats in the leading edge opened with a jerk which upset gun-sighting and must have let many a potential victim escape.

After trying a stall I happened to glance obliquely over my shoulder and was appalled to find the guns of a Spitfire. With a qualm I remembered that solitary hostile aircraft were making daylight raids, and that this Me.109 must have a most distinctive silhouette against the clouds.

I froze the controls rigidly as any pupil on his first flight—and flew very straight, wishing I was in cloud, blind-flying instruments or none. Whether the British markings super-imposed on the black swastikas showed, I had not the slightest idea. The seconds were hours—and still the Spitfire kept on my tail.

(I glanced again over my right shoulder. The machine was coming closer: another second and it was alongside with broad wing tucked between mine and the tail plane. The pilot grinned and jerked a thumb. Relying on the characteristics of a direct injection engine not to cut, I pulled back hard on the stick, and grinned to see the Spitfire shoot underneath as the Me.109 climbed vertically.)

(In an endeavour to retrieve his position the Spitfire pilot pulled straight up, but it only placed him in a position where I could make a short dive for his tail. I jammed the nose down so hard that a Spitfire or Hurricane doing the same manœuvre would have choked its engine, but the Me.109 did not even falter. I could hold the gunsights on the Spitfire for a moment only before he was off in a flick half-roll and a steep dive in the opposite direction.)

For a few minutes we climbed, twisted and dived after each other. Many times the steep attitude of the German machine evaded the Spitfire, and the abruptness with which the nose could be thrust down had undoubted advantage. (But in manœuvring at speed the heavy weight of the Me.109's lateral control proved most exhausting. Aileron could only be applied slowly and response was slow. I had to evade the Spitfire by gentle turns at very low speed, and try to match him with cunning.)

As I eased from a steep dive, using the trimmer to reduce the stick load, my base, which I had long given up as lost, showed unexpectedly in a cloud break, and I spiralled down. I slid back the stiff side window to see better in the murk, and a stream of ice-cold air swept in a still harsher engine note.

Slowly the hazy patchwork countryside wheeled past in endless pastels of dim brown and green, shapeless trees, roads, hedges, houses, slid under my squat wing. Like the turning pages of a picture-book misty field succeeded field, each growing larger as the machine dropped lower, until ahead the camouflaged sheds, the tarmac apron, and a line of Hurricanes took form. I passed over the near hedge, eased the stick back, checked, then dropped the tail a little more, feeling the machine begin to sink. Before I could attain maximum incidence the Messerschmitt was on the ground with the smooth contact of a 'wheeler.' The left wing losing lift quicker than the right, dropped a little as speed decreased: attitude steepened, and with barely perceptible jolt the tail wheel touched. I applied the brakes.

So that was that. One more strange aeroplane grown familiar—now no longer vicious, because it had borne me on confident wings through illusory skies which made the world a dream, back to the real earth where the skies were only faint remembrance. But the tail-high landing rankled. I taxied the machine around the perimeter, and presently took off again for a few experiments which would show the best method of making a perfect three-point contact. Then I parked the Hun near the Hurricanes.

Walking to the hangars through a drizzle of rain I glanced back at the British and German opponents standing on a muddy piece of land so lately saved for England by the Battle of Britain. I thought of their vapour trails twining through a cloudless sky: the far staccato of guns; a mounting score: the white billowing of a parachute canopy: black smoke belching: cattle grazing contentedly in sunny meadows: earth to earth and dust to dust. I remembered again the Immortal Few who had handled with such audacity those early Hurricanes and Spitfires—and I knew and gloried that even had they been equipped with Messerschmitts they still would have won the great and gallant victory.

CHAPTER XI

Lysander

It was a night of stars and moonlight. Beneath my aeroplane the silvered land pricked with the scattered lights of villages and towns, made navigation easy. The air was so smooth that the aeroplane flew itself. A great stillness haunted the heavens.

Presently my home aerodrome drew close, a large, dark patch encircled on three sides by the lights of the town. A burst of three green stars soared towards me, and I fired an answering red Very cartridge. Aircraft manufacturers did not yet use radio. For occasional night flying their pilots relied on dead reckoning and a flare path. But tonight for me there were no flares. I wanted to demonstrate that our new high-wing monoplane, the Lysander, could take care of itself.

I brought her round for the landing. The unique cantilever undercarriage was fixed, and the slats and flaps which automatically extended at slow speed also needed no thought. With a rumble of engine I placed her at 500 feet above the shadow of the aerodrome boundary. Speed was reduced and engine power increased to predetermined figures for a feather-like descent. More tail-plane angle was wound on until the Lysander was trimmed in such perfect poise that I could take my hands off the stick and let her make the glide and landing unaided.

Down she sank, hanging on the air rushing through the wing slots. In spite of the moon, visibility was too indistinct to gauge height accurately. Our stalled, blind descent felt far too fast and my confidence in the drawing-board figures and calculations waned rapidly as I waited, my hand poised to grab the stick, and hoped the undercarriage would withstand the impact with the ground.

But the calculations were correct. As the tail-wheel touched with a bump, I cut the engine, and the Lysander settled automatically on her main wheels. I let her run on until she stopped by the bright open doorway of the hangar.

None of us dreamed as we pushed her inside that within a few years the Lysander's destiny was to be hundreds of night landings in enemy-occupied territory, with no more ground aid than three faint hand torches.

I flew the first of the Lysanders in 1936 when we were still building the delightful little Hector biplanes, with Napier engines, as the standard Army Co-operation machine. The Hector, however, was an adaptation of the earlier Hawker Audax fighter, whereas the Lysander had been specially designed to meet military gun-spotting requirements.

Within eight years to the final spy-dropping version in 1944, the Lizzie was to grow more powerful and some two thousand pounds heavier; but she remained always the same queer, amazing machine—noisy, smelly, heavy on controls, but able to crump and bump in and out of an absurdly small space whatever the load.

The advent of the Lysander marked a new epoch at the Westland factory. Robert Bruce had retired and his assistant, W. E. W. Petter, a young Cambridge graduate who was later to achieve much renown, took over the reins. Everyone was anxious that his first design should be a success. He tackled it with brilliance. Unconventional in detail, the new aeroplane followed the logical sequence of Westland monoplanes that had been initiated by Arthur Davenport in the original 'parasol' Widgeon, followed by the Wizard, Witch, Wessex, and the P.V.7 that had caused my parachute jump.

Designed, built and flown within the record space of a year from the acceptance of the tender, the Lysander made immediate history. Ultra-modern it seemed—yet within a few years most pilots regarded it with tolerant affection as a peculiar old hack, long antiquated. That is the fate of most man-made things. The machine of today's specialist becomes the toy of tomorrow's youngster—soon to be discarded for something better.

I liked the Lizzie. Her methods were unorthodox, and though they resulted in no great speed or climb, her slow-flying characteristic was remarkable. No other aeroplane of similar power and wing-loading could operate from an area the size of a football pitch; but the Lysander's powerful slots and flaps made almost every small field a possible landing ground for emergency use. In a day when retractable under-carriages offering the refuge of belly-landings were yet to be, the Lizzie's capacity for a parachute-like descent gave flying a feeling of incredible safety. Because of this pilots would fly her in weather considered impossible for any other type in those days of limited ground control.

The convulsive rumble of the engine, the seeping of exhaust smoke from the depths of the cockpit, the rattle of windows and drumming of fabric sides, the ineffective jabbing of root slats at the propeller draught—these made up the familiar prelude to several thousand fascinating flights.

Take-off was simple. The only essential cockpit drill was to wind the large trimming wheel forward to the red mark on the scale. Failure could be fatal, for the tail-plane had to be set to an extreme negative angle when making even a normal landing. If the tail-plane was left in that position for take-off, the Lysander was apt to rear uncontrollably upwards, and several machines were wrecked as a result. This defect could not be readily overcome, and undelayed production was judged more important than the degree of risk to the pilot. Expediency has rationally governed such decisions countless times in the course of evolution of the world's aircraft. It is no good continuing development towards perfection for so long that the machine becomes obsolete before it is put into service.

Perhaps in these days the tail-plane would have been ingeniously interconnected with the flaps, or made to vary automatically with air pressure—but at that time it was simpler to insist that the tail trim was a vital check.

There was little else to worry about. The top knob of three on the dash was pushed in to obtain fine take-off setting for the two-pitch propeller. A stiff crank opened the cowling grilles.

As the Lysander rolled forward over the grass, the inboard and wingtip slats opened wide, automatically pulling down the flaps. The machine climbed noisily in a tail-down attitude at anything between sixty and eighty m.p.h., and there was always a strong temptation at first to hold the nose down to gain speed. To do so, however, meant that the slats and flap closed, reducing the lift and making the aeroplane feel as though it had lost all buoyancy.

A similar danger lurked on the landing approach. Speed had to be reduced boldly or a disconcerting longitudinal rocking could be accidentally started. As speed dropped, the slats and flaps opened, and a large backward movement of the centre of pressure on the wing sent down the nose, putting up the speed again and automatically closing the slats. To resume his glide, the pilot had to pull back on the elevator—only to repeat the sequence. The result was often a very untidy landing.

On a gusty day, too, the Lysander was liable to pitch convulsively as the slats snatched suddenly in and out. Coupled with the extreme heat of the cockpit and the exhaustion of moving heavy controls, this trait could be a little wearing.

The machine's qualities, however, more than made up for these foibles. Unlike most contemporary aircraft of that period the Lysander gave an unimpeded view in all directions. Perched high in the nose, with the engine at my feet, and a vast amount of window all around, I felt at first as though I was projected into space. England lay visible in more perfect detail than ever before, whilst I loitered with slats fully extended, exploiting the Lysander's slow flight.

Since those earlier days when the Widgeon revealed the intimacy of England, the country's features had become part of my deep consciousness. Only a glimpse through clouds was needed to recognize my position with fair certainty. The style of houses, their stone, or brick, or timbers, the size and shape of fields and woods, the flow of the land, the aspect of heights and mountains, all played their part in pinpointing my place in the familiar tapestry. Yet on no two days was the scene the same. The texture was constantly changed by the wayward moisture of the atmosphere, sometimes softening the landscape with mist; at others encrusting it with glowing light, or with a limpidity that painted the far hills with brilliant clarity.

But if the Lysander was good for lingering over peacetime scenes she was too slow for fighting and war, whose threatening undertones we already were hearing. Full out, at her rated altitude of 6,500 feet, the top speed was a bare 180 m.p.h. indicated. When pushed into a dive she became increasingly nose-heavy after 220 m.p.h., and with subsequent increase of speed it soon became necessary to wind the trim wheel back for recovery. At 300 m.p.h., control forces were almost beyond one's strength. At that limiting speed, the propeller screamed away far above its normal maximum because it lacked a constant speed unit. The structure vibrated with strain and the wing fabric between each rib bulged upwards under the tremendous suction.

In earlier tests, in fact, the fabric came off altogether. Although I had made many dives up to 340 m.p.h. with the prototype, even before an adjustable tail was fitted, the sister machine nearly came to grief when it was sent, after brief handling trials, to Martlesham Heath for official test. In a dive to 280 m.p.h., there was a sudden crack followed by a violent lurch. The Lysander plunged steeper. Squadron Leader Collins, the pilot, pulled back on the controls and the machine slowly responded. Glancing over his shoulders he saw tattered remnants of fabric beating in the wind, and gaping holes in the wing structure.

Cautiously he slowed the machine. The slats jerked open, the flaps came down. Soggy and unresponsive, the Lysander turned into a hurtling glide and Collins managed to reach the aerodrome and land at very high speed. The top surface of the wing had been stripped almost completely bare of fabric. Only the pilot's skill, the slats and wide leading edge of the spar had brought him home. Collins was awarded the A.F.C.

Suddenly the days of dreaming above a quiet England were wiped out. The long, golden time had closed. We were caught in a new web of life and death, and nothing could be done about it. Testing became a relentless task, where weapons must be rendered perfect for brave men. No more leisure. No more sailing for its peace. No more flying for its beauty and adventure.

And yet, even now, there was still the mind's escape: the impact of vastness and solitude: the consciousness of a power beyond oneself, which sprang not from fear of death, but from love of life that in those high places was stripped of the wrappings which shield it from earthly awareness of eternity. To all who flew, flying was an intensification of life. That was the reward of the many, whose wings for the last time reached towards the stars and nevermore returned.

Through those years of war the Lysander flew on. Her anachronistic role of spotting for guns in the strategy of outmoded battles changed to that of a defensive fighter with cannon in its wheel spats ready to rake the English beaches if an invader came. Relegated to target-towing she was subsequently reprieved and found a valuable and merciful mission as an Air-Sea Rescue aircraft. Her slow flying ability enabled dinghies and supplies to be dropped accurately, whilst fast motor launches were directed to the rescue.

Presently her characteristics of slow flying, short landing run, and quick take-off, came into their own. On moonlit and starry nights the black-painted Lysanders, fitted with a long-range tank beneath the undercarriage legs, made many a gallant flight from Bedfordshire into enemy-occupied countries. More than 300 agents were flown to France alone, and 500 brought back to England. Like moths the Lizzies travelled through the dim night by dead reckoning, to find somewhere in the dark landscape three electric battery-torches set in the form of a letter L, 150 yards long, by 50 yards broad.

A pre-arranged morse signal blinked. Slowly the Lysander groped down into whatever ground lay beyond the foot of the dim lights. In less than three minutes the machine would be off again and the field shrouded once more in darkness that hid a group of Frenchmen in the resistance movement welcoming a newly dropped agent.

The last I saw of the Lysanders was on a winter's evening in the final year of the war. In the darkened control tower we awaited the return of one of my friends in the spy-dropping squadron. We were a little anxious. For three hours the Lizzie had been humming through stormy darkness. Because the weather was so bad, it had been the only aircraft to fly from Great Britain that night.

Presently we heard the familiar note of its engine. Like a dynamic shadow it made a circuit under the low, dense cloud that hid the moon. No word passed because the single V.H.F. radio had failed. Recognition lights flickered and the Lysander emerged from the darkness on the broad band of the runway floodlight which had been quickly switched on.

I watched the silhouettes of two vague, shadowy figures climb stiffly from the rear cockpit and hurry into a car. At once it began to move and the noise of the engine speeding towards London faded into the silence of the night.

CHAPTER XII

Smoke of Battle

In a realm so high and vast and empty that it seemed the outskirts of eternity, I flew one day in 1943, dreaming my secret thoughts in a life maintained by an artificial stream of oxygen, and a pressurization system that warmed and compressed the cabin air to conditions equivalent to a relative height of 22,000 feet less than the true altitude. Perhaps because of its strangeness the hiss of the pressure pump struck louder than the rough growl of the wing engines.

At 47,000 feet I was so remote that the world of men had been forgotten long ago. Nor was I conscious that this Welkin prototype with a wing span of seventy feet was the biggest single-seater fighter which had ever flown. Instead I looked through the double thickness of perspex panels with a faint sense of uneasy equipoise. Wide wings projected over the glazed, endless sea which spread outwards until it flowed over the curved edge of the world and vanished into a silver void. Only through the narrow window of armour-tough glass in the front of my cabin could I find the strip of land that was the last of England: a reminder of the earth that I constantly forsook, but which still patiently proffered all that a man should want.

I stared with disbelief at the narrow, russet foot set like a fragment of amber in an enamelled sea. To man on earth it was the two-hundred-mile peninsula of Devon and Cornwall. I turned my face to the soothing warmth of the sun, ever the supreme reality in my infinity of space. As the inhabitants of any planet are physically unaware of its dizzy rush through the cosmic universe so I lived this remote moment of my life with no sense of the aeroplane's motion. Yet under the thrust of twin Rolls-Royce Merlins, the cabin in which I sat hurtled through the air at 400 m.p.h. Evidence of the swift path of flight lay visible in the mile long, billowing wake which the propellers were churning from the ice-smooth air. Such contrails were the new symbol of the age. Man could now change the aspect of the heavens with his passing.

A microphone was fitted in the oxygen mask strapped across my mouth but I flew in radio silence. I could listen, but, except for an occasional transmission to enable the flight plotters to fix my position, must not speak. Every few minutes terse instructions from ground to unseen flights of aeroplanes, from leader to pack, emphasized the dark, gigantic storm of war that pressed with slow deliberation towards the tiny, distant, tranquil island below.

Devon slid from sight. Cornwall shortened and narrowed into insignificance against the fullness of the sea. The last remnant of the land vanished, and only the pale shadow of the Scillies, far in the West, relieved the emptiness of the ocean. This moment marked the top of my climb. I pressed the stop-watch, registered the readings of the twenty dials ranged around me, and throttled back for the long glide to earth. The engines' note faded to a whispering quietness, leaving only the dull harshness of the compressor and the static of the radio to trouble my ears.

Occasionally through the mumble of voices difficult to understand, urgent words broke with startling clarity.

'Blue leader, are you all right?'

Silence—the skies waiting. Intently I scanned the emptiness: no sign of aeroplanes. Then:

'Blue leader—look out! Under your tail!'

'Christ—look out!'

'I think I've had it.'

'Quick, Mac. Try baling out.'

Silence again: perhaps the great silence of the end. Then presently more voices, different voices, staccato, unintelligible, threading the nothingness.

The black blades of the two propellers drew long, slow arches across the sky. The sun passed across the machine's orbit and hid behind the starboard wing. A further turn brought Cornwall materializing once more from the glittering ocean and looking like a shadow on the moon. The minute silver fissure near the sunward tip was the broad entrance of the Fal, and farther along, the scalloped shore reached to the vaguer indentation of Plymouth Haven. In form and beauty, slowly revealing the spirit of its shores, Cornwall grew to the sibilance of smooth wings descending through frozen air.

Whilst we were still very high, the bewitchment of extreme altitude passed, and the earth took on reality. Sombre hues of olive-green and russet separated into richer colours, and the sea grew edged with lacy foam. I could remember that men and women lived down there. I could sigh because the spark of their divinity was subdued by endless regimentation to which they tacitly subscribed, even though the whole wide earth was theirs. In my aeroplane I could dream I was of their race, yet free of human bondage. In unfettered flight lay a sweet illusion of immortality that could forget the ticking of the clock.

Ten minutes more and I would be seeking intimate contact with the land—flying low and close to the rocky shore, banking steeply round the headlands, skimming the glittering waves, at one with the wild host of birds my clamouring engines would stir into the air.

I opened power for a last high-speed run at height. . . . Noise rising in crescendo, and violent pressure on the rudder bar, startled me with the imminence of danger. I looked down instinctively at the small Cornish fields to see if there was anywhere to land, or what the chances were if I had to bale out. After the first second's uncontrollable desire to escape, a more disciplined mind took charge. I tugged back the throttle, but by then there was scarcely need to glance at the starboard propeller, shrieking in a blurred whirl whilst the other still rumbled round evenly and sedately. I checked with the dashboard instruments. The runaway propeller was 2,000 r.p.m. above the maximum permissible. Oil pressure and temperature were normal. I pressed the feathering button. No result: the propeller still whistled on at dangerous speed.

These things had happened before. There was no great cause for concern; flight had merely become a little more difficult. That was all. I knew that the source of the trouble must be the complex hydraulic mechanism which was no longer adjusting the swivelling blade angle to match the aeroplane's speed. Why it had failed hardly mattered, for there was nothing I could do except switch off the starboard engine and fly home on the port.

Base was 180 miles east. I reset the live throttle, dug my heel against the unsymmetrical pull on the rudder, and adjusted the bias control to counteract the load. Winding the handle was arduous work and brought black and grey dots before my eyes. Inside the cabin there was unexpected quiet. I realized dimly that the compressor must have failed when the engine raced away. Bereft of pressure, the atmosphere I breathed had changed without warning from the equivalent of 7,000 feet to a true height where life could last only a few more minutes. I groped for the oxygen regulator and turned it to maximum flow. Next moment I felt normal.

It was then that the ominous ripple of oil began to spread

from the propeller spinner across the engine cowl. Another flying glance at the instruments showed the oil pressure was very low. With the engine windmilled uncontrollably by the propeller, bearings would quickly heat and seize if the oil escaped too fast. An immediate landing was imperative.

I called up base. 'This is Owlbird. Angels twenty-five. Pancaking at nearest aerodrome. Zero oil starboard. Impossible to feather.'

There was no answer.

'Owlbird calling.'

Still no answer.

'Owlbird calling.'

Nothing. Radio silence. More important things were happening. We were at war.

My gaze surveyed the whole peninsula. Almost every part was within easy gliding distance, and if I could but locate them, at least three large aerodromes with long enough runways for a safe landing were hidden amongst the fretted contours of the shore.

I trimmed the machine into a fast glide that lost half a mile of height in sixty seconds, and would take five minutes to reach the earth. Five minutes: too long, perhaps, for the ailing engine. I needed a little luck that would guide me in one swoop to an aerodrome as yet hidden from my view.

Land's End orientated me in plenty of time. From 10,000 feet, the harbours of Newlyn, Mousehole, and Penzance took recognizable shape, and Looe Pool, where the sea notched deepest into the land, gave a friendly glitter. The familiarity of an old story was everywhere. By the side of the Pool an ancient wood encircled the park and the grey house where long ago my forebears lived. That bird-haunted water, where those ancestors found quiet recreation, pointed my way of escape. Only a mile away was the aerodrome I needed. It came in sight as I banked the aeroplane, and in two minutes my wheels were touching down on the runway.

Later that afternoon, I started the journey back—but not

by air. In a borrowed dinghy I drifted on the ebb from Gweek, at the head of the Helford estuary. Friends had arranged that on the next tide they would collect the boat from the Ferry Inn downstream, leaving me free to make my way by road to Falmouth so that I might catch the night train home; but first for an hour or two I could re-live an old enchantment.

To round the bend beyond the crumbling quays of Gweek was to enter a scene that knew nothing of modern ways. It was the waterside of a hundred years ago: solitude and an arc of sky, and the broad river flowing seaward between high, sloping ground where trees crowded to the water's edge. Except for a cottage or two, almost hidden at the mouth of the Mawgan creek, it was a world of flighting waders, and soaring gulls. Grey herons waited in the shallows to spear fish, and buzzards circled on upswept wings among a host of jackdaws and rooks above the windward bank—and high above them all the dark anchor shape of a peregrine lorded the river.

No need to follow the twisting channel for the tide still held, covering the mud so deeply that it was possible to row close to the banks beneath the overhanging trees in sheltered tranquillity. The brooding quietness was infinitely soothing, assuaging the empty questioning of the mind, lulling memory of noise and hate and turmoil, exorcising doubt to find the foundation of old hopes. Thought became gentle-winged, soaring in facile circles, like the sun-bathed pinions of the gulls flashing white in the cloudless sky.

A call of sudden alarm from the curlews at the sight of the boat drifting round a spur of rock emphasized the silence. The birds flew swiftly upstream, their fading twin-tone notes echoing from a high, turfy slope marked by the straight vallums of a Roman camp. With what emotion, I wondered, had those bronze-armoured, long-dead men listened to that same call, or the guttural *kronk*, in the long, evening shadows, of herons nesting in the branches below. And, before their time, the swarthy, blue-daubed Britons had known the same river sounds as they edged downstream in their clay-caulked coracles to meet Mediterranean long-ships waiting in the deeper water below Polwheveral creek to trade for tin.

With lazy strokes I rounded Groyne Point and took the boat a little up the creek—far enough to make the bend and see the water stretching a crooked, silver finger into the distance between clustering trees and a down-like sweep of pasture. Beyond, round the bend, the two banks closed to a narrow strip where a copse half hid an ancient water-mill. Long ago tides had worked its mossy wheel, and the mud-flats had muffled the dull rumbling of creaking machinery; but now the air rang to the whistle of oyster-catchers and the scuttering cries of small wading birds.

No time to rediscover it today. I turned the dinghy downstream. As the bows swung round I saw the craters of a stick of bombs—scars of modern civilization to endure, despite 2,000 years of ploughing, like the ditches of ancient man.

I rowed slowly. The river banks spread wider. Beneath the flat shore of Lower Calamansack I spied a friend of old—a black and yellow boat, whose crew were dredging oysters. On the other shore the fringe of oaks ended, and Frenchman's Pill was revealed—a shy, romantic creek. On its farther bank splashes of rambler roses marked terraces among the heather and bracken. By the little quay *Tar Baby* floated patiently, as of old.

Like a broad lake now, the water flowed wide and shimmering between low, rocky walls. For a mile seaward of the Pill, heather-covered slopes were banked with pines, and dwarf oaks clung to the rocky shore. The opposite bank was bathed in sunshine, and though it was early autumn, the soft air breathed summer on the multi-tinted green and bronze, where stubble wheat was hedged from cropped meadowland, and the fields climbed gently to a sweeping line of dark tree-tops which led to the secluded anchorage of Abraham's Bosom.

A few energetic tugs of the left oar carried the dinghy across the river towards the still hidden entrance of Navas Creek.

Herring gulls screamed and plunged among a shoal of white-bait, and further inshore, sudden splashes and a taut blue curve told of mackerel hunting the same quarry.

Reaching the mouth of the creek, I looked downstream and contemplated for a moment the fully matured estuary. It was unchanged: serene and quiet, dreaming in the sunlight. The old charm of the immemorial waterway was a cloak of assurance: a promise that the lost things, the cherished, had never really gone. In the soft Cornish sunshine, the spirit of that place waited with tranquillity, listening to the liquid call of the birds, content to hear the sighing wind and the soft swish and tinkle of water thrown from a yacht's bows. Ghost after beautiful ghost I saw, tugging lightly at their chains—all the gallant company of lovely ships which the unfolding years had brought to the magic of that haven.

I drifted on. The fairway turned for the last time. Now I could see the entire expanse of estuary spreading far away to Rosemullion and Nare, and seeking the final freedom of open sea and sky. Three motor fishing vessels in the roadstead were silhouetted against a splash of pale turquoise ocean.

Remembering many things, hearing the far echo of music, laughter and talk from bygone ages, I rowed the last few yards. A desolate, mewling cry came down from the sky. I looked up. A speck in the blue, a buzzard soared on motionless wings, at liberty to fly all the days of its life, in sun and in storm. Meanwhile mankind exchanged barbarism for barbarity, in the struggle to flee from the wilderness. Must I excuse it as destiny because his restlessly questioning mind was born to fight as well as dream? The quiet flood gave no answer, but continued its even, passionless flow to sea.

High overhead the buzzard cried again—and I remembered with relief that I too had wings.

CHAPTER XIII

Meteor in the New Dawn

In the form of each successive aeroplane lay the seed of the
next, although rarely visible until some years later. It had
always been that way—from the first fragile wood and fabric
creations to the all-metal shells which solved the requirements
of utmost strength and smoothest shape. Fighters had grown
as heavy as the twin-engined bombers of the previous war.
Throughout the following decade they were to increase in size
and structural density until they reached the weight of the
four-engined bombers of the 1939–45 war. But first a revolu-
tionary change in type of engine had to pave the way to a
degree of power and performance inconceivable in the idyllic
days when first I flew.

'It's only the beginning but the implications are terrific,'
Gerald Sayer had said to me after he made the first flight in
1941 in the Gloster–Whittle E28 machine designed for the
earliest British jet engine.

That first test flight had been a tremendous leap in the dark.
However dispassionately planned, it required not only a

brilliant pilot but a man of courage, who could observe, analyse, and report facts clearly, whatever hazards might be encountered. All those qualities were Gerry's as well as his charm. He would have smiled deprecatingly, were these words not an epitaph, for a year or so later he was killed in a collision while flying with an R.A.F. formation.

By that time such a wealth of information had been accumulated from his flights on the first single unit machine that the way was open for the next—a twin-engined single-seater fighter. Inevitably it suffered teething trouble due to its novelty and speed. Seldom has a new design evolved without much alteration and adjustment, and every advance takes a toll that often seems intolerable. The Meteor was no exception. It exacted a quota of lives from both civilian and service test-pilots, but from these sacrifices emerged a perfected, vital weapon, that became the fastest fighter of the post-war decade.

In the summer of 1945 I was given the opportunity of flying a Meteor from the first R.A.F. squadron equipped with jet aircraft. Except for the canvas covers shrouding the jet nacelles to keep out dust, there was no secrecy about the cluster of strange new fighters at each dispersal point. So low were the tricycle undercarriages that at first impression the machines looked as though they had belly-landed.

As I walked to my machine, twelve others in close succession reached the runway and formed into neat formations of three. The noses of the first flight bowed as wheels strained against locked brakes under the thrust of the turbines and then the three machines moved smoothly forward, holding as tight a formation as any of the stars of pre-war R.A.F. displays. As soon as they had travelled a bare hundred yards the second trio began their take-off run, followed at similar intervals by the third and fourth flights. The tarmac shimmered with the heat of the jets skimming the runway, but the turbulent air in no way affected the precision of the formation.

A routine squadron manœuvre, the massed take-off was an impressive demonstration of the confidence these young pilots had in this new fighter with its novel power plant. Certainly it tempered the customary caution with which I approached the flying of any strange aeroplane.

The light, spacious cockpit in no way differed from any conventional fighter, except for the superb view. The seat was far forward and the wings and short projection of the turbine units gave negligible interruption in the vision. Coupled with the tail-high position afforded by the tricycle undercarriage, the steep slope of the cowling enabled the ground immediately in front of the nose to be seen.

The squadron commander briefed me on the controls. Bolt-like handles sliding in long guides, one above the other, operated the turbine thrust. Pairs of co-axial high and low pressure cocks, set to port and starboard, looked sufficiently like any normal fuel control system to be readily accepted substitutes in the cockpit drill, and the balance cocks interconnecting the tanks were typical of most twin-engined aircraft.

Only on looking closely at the instrument panel was there evidence that this machine was something new. Gauges in a prominent position showed not only the burner pressure with the danger point of 520 lb. per square inch, marked in red, but fantastically high figures calibrated well beyond the working limit of 690 degrees C. for exhaust temperatures. Revolution counters on the top of the dash seemed to have added a nought too many to calibrations reading up to 18,000 per minute. Yet the small, barrel-like swellings on the wings which housed the power-plants looked so inanimate and harmless.

I ran through the starting procedure: high pressure cock down; low pressure O.K; balance cocks up; trim tabs neutral; switch on pressure head and D.R. compass; pull throttles back; see the starting crew are ready . . . switch on port supercharger.

The red fuel light glowed and I pressed the starter button, drawing a gentle rumbling from the jet unit. The needle of the rev. indicator moved round swiftly. At 2,000 r.p.m., I started the starboard unit and within a few seconds both were idling at 5,000 r.p.m. I looked at the gauge unbelievingly. It seemed incredible that anything could be turning so fast, for there was no vibration—only a dull noise from the jet pipes, and the impression of a heavy top spinning effortlessly.

No waiting: no warming up: the Meteor was ready for taxying. A cautionary glance at the brake pressure, and I waved away the chocks and shut the hood.

I opened the power controls a little. With the slight pitching motion of a speedboat the fighter moved on to the track, steering easily on the brakes. I found no strangeness in the absence of propellers. It was enough that the machine moved forward at a touch of the throttle, and within a few yards I was in tune with the aeroplane. I began to taxi faster, following the curving track by differential use of the throttles. There was all the steadiness and control of a car.

When we had moved to within a few yards of the runway the control officer flashed his red Aldis lamp. As the machine came to a stop one of the squadron came into the approach too high. Realizing he was going to overshoot the pilot opened up and the trail of the exhaust puffed and thickened. Without hesitation the machine climbed away.

'Good,' I thought. 'The turbine pick-up must be excellent.'

The green Aldis winked. I taxied on to the runway, and held the Meteor on the brakes. Flaps set at take-off, a last glance at the instruments and I opened both throttles steadily. The machine squatted and strained as the nose oleo compressed. The exhaust note changed tune. The revs soared up: 10,000, 12,000, 14,000, 16,000. I released the brakes and the Meteor ran forward with a pleasant acceleration, so that again I thought how like a car it felt: an impression emphasized by the wide ribbon of black tarmac runway flowing swiftly past the nose.

A strange quietness in the cabin made it difficult for me to feel the vital moment for take-off. The seconds raced past in a silent surge. The long runway rapidly dwindled, but still the machine seemed heavily planted in the ground. My eye sought the A.S.I. but instead of the 80 m.p.h. needed for take-off it registered '16'. An incredulous second elapsed before it became clear that my hasty glance had fallen on the most obvious dials, mounted as they were above the rest. The '16' was 16,000 r.p.m. that the turbines were giving. An instant later I found the A.S.I. The speed had reached 120 m.p.h., and a bare quarter of the runway left. The opiate of fatalism dulled my mind to a queer indifference that if the machine stayed on the ground any longer it would crash through the aerodrome boundary and burst into flames.

I pulled on the control, and the Meteor responded effort-lessly, leaving the runway and slanting upwards against the air. A slight shuddering reminded me to retract the under-carriage, and flight settled into a perfect smoothness. There was not a tremor. Except for the steadily mounting hand of the altimeter the instruments seemed frozen into an unearthly still-ness. Already we were at 5,000 feet. I levelled the machine and throttled a little. The noise increased slightly, but it was still small as the pointer of the A.S.I. settled at a level cruising speed nearly as fast as the maximum authorized diving speed of the Spitfire. The Meteor was streaking through the sky in a manner which left me in no doubt of her progress. Within five minutes half a county was crossed and left far behind. For the first time in all my flying, I gained an exhilarating sense of speed. In one hour the Meteor could cover the same distance which my light aeroplane of twenty years ago took a day of flying and re-fuelling to complete. Here was the magic carpet at last.

Think of a place, think of anywhere in the length and breadth of these isles, and before thought would grow tired of journeying we could be there. Not only England, or Europe, but all the world soon would be within scope of a few hours' dream-swift flight.

Gone was the imprisonment of those engulfing chasms of sound which hemmed the solitude with deafening silence.

There was only whispering quietness, fainter than the sigh of a breeze, as though the timeless waves of eternity were rustling against the sides. In this spartan metal shell, crammed with a thin bubble of transparency, life attained a monastic simplicity, the mind calm and contemplative beneath the vastness of unending space. Only the present remained, though I could see signs of the past and feel the flow of time drifting endlessly into the future. I could fly forever on these dream-wings, still and silent above the misty panorama that must be the world I knew in some other life than this.

I glanced at the fuel gauge. The kerosene was flowing to the burners fast as water from a tap, and the tank was already half empty. Time was running out. I must forget the magic. I pulled the machine into a climb that took me twice as high in a handful of minutes; then from 40,000 feet tumbled down again in the exultant curves of three-dimensional flight. There was no doubt about it: this fighter was a lovely aeroplane to fly. It had all those indefinable qualities which bolster confidence; steadiness, a sweeping forward view, ease of control, swift and sure response. Despite its great speed and weight it was unquestionably an easier and safer machine for any pupil pilot than the last of the piston-propeller fighters, endearing though they were for their exciting manœuvrability. Engrossed in assessing the technical merits of the machine I forget the passing of the minutes until it was time to be anxious about the fuel. I turned for base, thirty miles away, and had no sooner settled on course than the three black intersecting lines of the runways appeared ahead.

I throttled back, but so clean were its lines and so great the speed that the Meteor sailed on effortlessly. Before the machine could decelerate we had swept over the aerodrome, and a couple of wide circuits were needed to bring the A.S.I. down to the preliminary approach speed of 200 m.p.h. I moved the flap a quarter, and the Meteor responded with more conventional behaviour. At an indicated air speed of 150 m.p.h., I lowered the flaps fully, arresting the momentum and bringing a moment of needless anxiety that the turbines might have stopped. The uncanny silence emphasized by the absence of fanning propellers, was typical of a dead engine, but a glance at the instruments and the instant response to a touch of throttle reassured me of the effectiveness of the invisible source of thrust.

I selected the undercarriage levers for the down position. The red lights glowed; the starboard turned green; but the port and nose lights remained a warning red, though the mechanical indicator of the nose-wheel had moved. Opening up the jets I waited a long minute for the two lights to change. They stayed at danger.

I pulled the lever up, and the lights quickly faded, indicating that the undercarriage had once more retracted and was locked. Evidently the pump was functioning satisfactorily; but when I again tried to lower the undercarriage only the starboard wheels signalled that they were down.

For the second time round I passed the flight buildings and the black cross of the runways scarring the broad turf; then no more aerodrome—only small, tree-enclosed fields. The tanks were nearly empty. Maybe I had seven minutes of flying left. I opened up to gain more height and find space enough to think.

I retracted the wheels into the wings. Red lights—none. Hopefully, carefully, I selected the down position. Red lights for port and nose, one green light for starboard—one leg up, one down, and no forward wheel. And I was making a first flight with a borrowed machine! Caught with no radio contact, I was unable to ask Control for a visual check on the legs.

'Use the emergency hand-pump until it goes stiff,' the C.O. had answered my question about procedure in the event of undercarriage failure.

Changing hands on the control column, I found the long lever of the undercarriage emergency gear projecting through the floor on the starboard side, and began to move it vigorously backwards and forwards. I pumped and pumped. The two red lights still glowed. I pumped again until I was sweating and breathless, but the lights did not change. I conceived an intense dislike for my aeroplane. Fuel was almost gone, and the perspiration running down my face came not only from working the pump. In a minute or two a landing would be inevitable and then?

We made another circuit of the aerodrome. At the end of the operational runway stood the striped control van from which the Aldis lamp had flashed permission to take off. Surely they would flash me a red signal if one of the legs were up. If they did, I must retract the other quickly and make an ignominious 'belly' landing. I stopped pumping. Too little fuel was left for any other decision.

I brought the Meteor round in a wide, semi-power circuit, and straightened two miles away for the approach. At a little over 110 A.S.I. the machine glided towards the runway. I watched the control van intently. No signal. The boundary markers slid underneath, and the stretch of grass between machine and runway rapidly diminished. The control officer was watching. I saw him look casually away. Now what? . . . Only a ribbon of runway stretching ahead, and the Meteor's stubby wings a few feet above it.

I eased lightly back on the control: the tail dropped lower, and with a squeal of tyres the aircraft was racing along the runway, poised at a slight angle on its main wheels. Slowly it nosed down. A second's suspense followed whilst I waited for the port leg to collapse. Instead there was a slight jar and the nose-wheel grounded. A touch of the brakes, and the Meteor checked its run, stopping easily in half the runway length.

I taxied to the perimeter track, and back to the dispersal point. The tank gauge showed empty. My face was running with sweat.

The moment the jets stopped, the C.O. climbed on the wing. I opened the hood.

'Well?' he asked, and I gave him the answer he wanted.

'Terrific.'

'Perfect!' he said.

I nodded.

So opened the portals of a new threshold and a different future. Here was the power and the efficiency of a mighty propulsion. Already the scientists and mathematicians were thinking of the razor-thin wings and arrow shapes that would overcome the aerodynamic barrier of the compressibility shock-wave—the next hurdle to unlimited speed. Swifter, bigger, more complex, more efficient, capable of still greater range and heavier loads—these aeroplanes would grow into the practicability of a designer's logical visioning, and the reality of massive metal structures fabricated by the co-ordination of many hands.

CHAPTER XIV

Chariot of Fire

The year was 1952, but that meant nothing up there except a cipher with which to use the star tables. In the faintly throbbing silence of the flight deck each sat wrapped in thought, subconsciously dealing with the routine of flying a big four-engined turbine-jet aeroplane at stratospheric heights, half across the world and back. Through the courtesy of De Havilland's I had earlier been allowed a spell at the controls of the Comet, and I had eagerly accepted the invitation to fly on the BOAC familiarization flight to Singapore and back. Enclosed by matt black walls which were patterned by a hundred dark dials, we peered outward past the windscreen, as though from a tunnel's mouth, into the sunlight beyond.

The vast dome of unimaginable heights was brushed with black and set upon a wall of silver-frosted blue. Its base formed the remote horizon encircling the microcosm earth lying frail as a fallen leaf upon the oceanic surface of the world.

In spite of the epoch-making swiftness of our flight, height retained the old yet ever new illusion that the aeroplane was poised unmoving in high space, whilst hour by hour the sun's white blaze crossed its track, brushing cloud and sea and land with endless combinations of light and shade. Confirmation that even in the frozen stillness a scale of time should measure distance, came from the speeding minute hand on the dashboard, though ten minutes or an hour of flight could equally well embrace the whole eyespan of space between wilderness and man's salvation.

Indeed it was the wilderness, the desolation, the primitive upheaval, the ravaged tiredness of the world drifting slowly below that outweighed all other impression—that and the implacable, overwhelming vastness of the sea. Yet in the skies above, transcending anything on earth, the great stillness brooded, extending upwards and outwards and all about, like a cloud of incense above a brazier, flooding through countless aeons of time yet finding neither end nor beginning.

Through the gauzy curtains of space we stared down at the great areas of jagged mountain tops, and rolling forests; marshy deltas and jungle river courses; pink-tinted deserts of barren sand. But whatever the desolation, fingers of cultivation plucked insecurely at the fringes of the primitive and untamed. The vestiges of earth that man had laboriously smoothed in the thin crust covering the primeval rock were pitifully clear to see: startling evidence that of all the world's land no more than a tenth was man's heritage wherein to feed 2,000,000,000 inconsequently begetting and multiplying people who dreamed the earth was made for them.

Down below, misted and beautiful, was a world of fantasy, an idiot's dream, a cataclysmic upheaval frozen in its agony, bathed with radiance and painted with delight. Slowly as a cloud travelling across the path of the sun, it passed from sight, and instead there was only the emptiness of sea. Between endless water and unending space we stayed suspended, inanimate, content to wait in the warm, radiant whiteness of the sun, as though at last the deepest secret had been revealed and there was nothing more to seek.

The faint vibration of the whirling turbines buried in the wings flooded like a thin stream through the metallic fibres of the structure. The pulse of its own life beat with ours as the aeroplane persistently aspired to horizons hidden beyond the rim of the world. Presently a purple shadow on the ocean became another land, its roots hidden in the depths of the great waters that are the essence of this world. Soon there was no more sea, and we flew, unimagined and unseen through the high space of yet another continent.

How could I guess that the great stillness was illusion? How could I know that the timeless plain, the ageless desert, the smooth ice-cap, the snow-topped mountain, the unstirring chasm, or the slumbrous island floating on an azure sea, had none of them known quietness? Yet from this great and revealing height, there were many signs, if they could be interpreted, to show that the bones of the planet were not petrified but living—swelling, contracting, shuddering, moving with sluggish pulse. Slowly, infinitely slowly, new mountains were being raised by this dynamic energy, and old ones were being eroded by the rough hand of time. Fissure and faults were forming under the strain of unbearable internal stresses. New lands were appearing, and long-established shores dwindling. Sun, rain, and frost were eternally smoothing the ruckling, heaving skin of sand, chalk, clay, and decay that covered sedimentary rocks long disrupted from their fundamental creation of granite and basalt which encircled the nickel core of our planet. It was a surface world of inexorable and ceaseless change—an earth still tortured, still tormented, built on an unending pattern of destruction and rebirth.

Hour after hour, riding the thin, unbreathable air in the

deep suspense of space, we stared from the artificial life of the pressurized cabin at the gradual unfolding of thousand upon thousand miles of natural existence. Only at rare intervals did the high turbulence of great air currents sweeping the skies of the world interrupt the smooth steadiness of the flight. Otherwise the absence of propellers added to the illusion of timeless trance in which the unrolling panorama of a world was displayed in all its dwarfed nakedness.

Through the armoured side-windows I could see the pointed tips of the tapered wings far astern of the long nose in which we sat. The air intakes of the four jet turbines were hidden by the streamlining of the cabin wall, so that there was no sign of motive power, and the aeroplane seemed to be floating divinely on the high winds of heaven.

There we sat, four of us, waiting, warm and at ease in the sunlight streaming through the windows, aloof with thoughts that were our web of life. With unhurried mind we watched the stark loneliness of untamable places. Nothing could hide the fact that despite his fertile land, the life of man hangs by a thread in a world of chance where it matters nothing whether he survives or dies.

On and on we flew, in sunlight and in cloud, by day and by night. Every few hours we descended swiftly to earth, and resumed mortality, half suffocating in the tropic heat, while the great fuel tanks were quickly filled. Then once more into the frozen heights of the upper air, that had to be compressed from thin iciness to burning density, cooled by refrigeration and made humid with moisture to suit our frail lungs.

The drop of the sun below the edge of the land brought night unheralded; and in the tropic darkness the stars by which we checked our flight were pin-points, no longer beautiful with friendly loveliness, nor intimate like moon and earth, but coldly geometric.

Time waited—until dawn caught our flight, a quick flame flickering in the dusky curtain of the east, and draining the sky to cold, bleached blue. Scattered clouds glowed, grew firm and white, and with a single breath the universe threw off the last veil of darkness. Across the edge of space, the horizon cleared and the rim of the sun flooded the earth with daylight.

Always it was the same—and always different: the white edged, burning blue encircling the horizon; the powdered black of outer space; the remote cloud floor, sometimes stretching over the world like a level snowfield, sometimes thinning and separating into wisps and billows, and revealing between deep rifts the sober colours of forgotten land. Cloud shape changed in form and texture. Smooth, level plateaux tinted with soft gold and blue, broke into towering cumulus of breath-taking grandeur. Diffused veils of smoky strata spread in vertical layers a mile apart. Black with the violence of internal turbulence, gigantic anvil shapes of cumulo-nimbus resisted our flight with storm and lightning. Everywhere and in every shape clouds signalled the stir and circulation of the atmosphere, and expressed the world-wide cycles of dynamic, omnipotent movement, bearing the water of life to what would otherwise be dead dust.

Far away on earth man might feel the wind and rain, and cower for shelter from the darkening storm that hid the landscape. Viewed from the stratosphere that same storm might be seen shadowing a hundred miles of land beneath white-quilted cloud that could be covered with two fingers. Beyond lay pools of sunshine within reach of a mere fifteen minutes of flying time.

Hours became days whilst we watched new land and sea and sky. A silver insect trapped beneath the domed ceiling of high heaven, our aeroplane drew its evanescent trail swiftly around the earth's circumference. To us it was the very matrix of the universe, the living womb cradling our lives, enclosing crew and passengers protectively with warmth and infinite assurance. Its cabin, trembling faintly, was the only reality in which we could believe. A score of races of mankind must have heard the faint singing of the unseen passage of our stratospheric flight. To them, what could it matter, that a jet airliner heralded another epoch in the history of the earth? The sun and rain, the plots they tilled, their hunger, and their love remained unchanged.

Yet the voice of man continually reached up to us with speech and signal, for we rode the path of wireless waves guiding the long trajectory of flight to the hidden target of each further destination. In a long line forming the image of the earth-track of our skyway, station after station listened for the call sign; heard; reported our bearing; and passed us on. Beacon and track guide, station fix and radar cover, sent vibrations trembling through the air, to be transformed through electronic valves into speech or staccato note or image that linked us to the care of men on earth. But it was a frail thread. The clouds had only to raise their electrically charged heads to distort with static and perjure the signals on which we placed our trust. Even the softness of night could ruffle the rhythm of radio waves into a false echo of their true intent. It was then that the message of the stars speeded our course with certainty as though they were the lighted pinpoints of a map. But when presently from star-fixed space the long descent to earth must be made through the dense opaqueness of many layers of cloud, we flew with tense alertness until radio aids once more reached us without distortion.

Shrouded in blinding mists, the illusory senses of the body relied for balance and direction upon the readings of a multitude of dials. No aerodrome could be located by act of faith in compass and calculation where an error of ten miles was high precision. In clear sunlight, or even unclouded darkness, landmarks by day and lights by night added visual direction to our destination. But of all our journey endings, full half were made in mist and rain, or with cloud so low above the earth that the restricted space of air between seemed suffocating under the hugeness of our wings. It was then that the guiding signals of radar and ground control were a matter of life and death.

In the tenth hour of the last day's flight the radio led to the final long let-down from high above a cloud-covered France. Within ten minutes a ripple of water gleamed through the mist-blue distance—and suddenly the endearing smallness of southern England grew out of the horizon as though fanned to flame from the spark of faith in our minds. Only a strip of sea and a diminishing gulf of air remained between us. The shores of France slid past the cabin windows and were gone with a seeming whisper that it was not the empty deserts, the lifeless mountains, nor the desolate seas which separated man from man, but the wilderness in human minds and hearts.

Expectantly we descended towards our homes and loves. England was suffused with the level light of evening. Memory of the world's barrenness in ocean wastes and tortured nakedness of land vanished as though we had never speculated on its cruelty. The echo of primitive fear of the earth had died. The impact of vastness and solitude was forgotten. The awareness of the power throbbing through the universe faded. In their train came a sense of relaxation and infinite peace under the spell of England's charm: her gracious sweep of downs; her quiet vales of trim meadows and neat woods; her inimitable quality of greatness and power.

Soon we would land, the long runway hissing under the wheels, the machine settling its weight more and more firmly on the ground until at last it was wholly sustained. I would step again upon the good earth, look about me, breathe the evening air. The rich reality of all the sights and sounds of mankind I had forsaken for the splendid solitude of flight, would return. Presently there would be lamp-light—the candle-flame of life and love, the murmur of voices; the shelter and closeness when darkness would be a velvet curtain of privacy; and the stars sparkling jewels devoid of all significance except their beauty.

CHAPTER XV

The Bell Tolls

Like a procession of great cumulus clouds, painted with sunlight and shadows, year after year had sailed into the past. Looking back it seemed a far cry to the day when flying first enthralled me with wings of canvas stretched tight with dope upon a wooden frame of spindled spars and latticed ribs, and propelled by a stuttering rotary engine. Step by step aeroplanes had become more mechanically purposeful and complicated. The bird-like purity of control of machines with small inertia had gone, and we accepted without further thought the ponderous flight of heavily laden wings drawn by fantastic power at speeds which annihilated time and distance.

The Wyvern, a strike fighter, which I tested through difficult years between its initial flight in 1949 and introduction to Squadrons in 1953, was the three hundred and ninetieth type I had flown. It had huge, turbine-driven, contra-rotating double propellers which gave a misleading air of anachronism among the rapier-pointed jets. With wing area little greater than my light aeroplane of old, it was heavier than the twin-engined transports which only five years earlier had carried detachments of fully armed parachute troops to an invasion designed to be the end of war. Yet in the succeeding days of peace man lived more uneasily than ever before, filled with foreboding. So my task remained the same as in the beginning: developing and making safe for other pilots ever newer forms of airborne death; and finding in that strange pursuit fulfilment of the mind's most eager striving, because it let me see the world, the universe, and men in their most profound relationship—and in so doing let me find myself.

This time there were no vast heights to scale. The measure of progress was that the Wyvern could lift four times the weight twice as high as a similar aeroplane of twenty years before. Its speed did not approach that of the latest jet-reaction fighters which could already escape the sound of their passing, but it flew faster than the swiftest fighter of the 1939–45 war, and nearly five times the speed of the little Widgeon monoplane of those long ago days in the dawning age of flight.

From the beginning when the engine of the Mark I proto-type stopped suddenly as a result of ignition failure 20,000 feet above a low layer of stratus cloud, it sharpened our wits. The cause of that first forced landing was comparatively un-important, but, as always with the development flying of the Wyvern, it was the effect which was grave. Although the all-up weight corresponded to that of a Dakota, the wing-loading, more than 60 lb. per square foot, was nearly three times as heavy. Consequently when the engine cut, the Wyvern dropped, more like a brick than an aeroplane, in a fast, steep dive because it became impossible to feather the propeller; and the enor-mous drag created by the large, flat, paddle-shaped blades in original use, made it very difficult to flatten out for landing.

On that occasion I hurtled through the cloud base at 3,000 feet to find the disused aerodrome of Warmwell luckily within reach. Miraculously the machine landed safely between hidden concrete blocks that had been scattered over the entire overgrown area to dissuade German airborne invaders.

The first eventful circuit of the later prototype powered by the gas-turbine replacing the Rolls-Royce piston engine, was the briefest initial flight I have ever made. It lasted barely three minutes. At that moment of becoming airborne, the cabin filled with dense, acrid smoke. I had to turn on full oxygen to breathe, and landed unable to see the instruments and convinced the machine was on fire.

Yet there was no real cause for anxiety. The fumes had escaped from a hydraulic leak pouring on to the hot case covering the incandescent gases behind the combustion chamber.

The main problem of the Wyvern's long and laborious development arose, however, from the installation of the turbo-prop power unit. Turbine thrust was adjusted by control of the propeller pitch which could be changed as quickly as the pilot moved the throttle, enabling thrust to be cut sharply for landing on an aircraft carrier. Whilst satisfying the deck landing requirements, the rapid pitch change of the propeller was at high speeds inclined to release the energy stored in the rotating parts of the engine, and bring about violent surges in the thrust of the turbine.

In proving the solution a colleague and I did a long series of dives to see if the engine/propeller speed control was liable to run away and cause serious damage. With each pull-out subjecting us to 3G or 4G, we were very tired by the time we had accomplished a thousand dives; but the investigation established the reliability of the control unit and led the way to endorsive tests by Naval test pilots.

Before this there came a particular moment in the sequence of testing when, my job complete, I had prolonged the flight to throw circle after circle in the clear sky of late summer afternoon, loath to forsake the high view of Wessex drowsing in sunlight. Presently I throttled the turbine, and the tandem propellers on their single shaft hummed in a long, straight glide from three miles high to the home aerodrome.

At 800 feet I opened the engine to a livelier tune, and started a gentle turn to bring the aeroplane round for the approach circuit. With no imminence of danger, disaster struck wildly, flinging the machine into a violent sideways swirl. The sky spun round, no longer over my head, but under my feet. Ten times swifter than normal thought, subconscious animal instinct had swept the control column across the cockpit in the opposite direction to the wild sideways spin. The ailerons were ominously weightless, giving no perceptible response; yet the rotation slowed and I was left hanging inverted in the straps, diving upside down at rows of neat red houses set with the distortion of a dream above my head.

I could see no escape. The machine was uncontrollable and those wedge-shaped roofs were already barely 500 feet away. The seconds were luminous and still. A fraction longer of living and then walls would crash asunder as the aeroplane shattered into fragments under a pall of flame and oily smoke. From far away I watched my end dispassionately.

Yet swiftly in the fleeting instant between life and death the ponderous machine was doing strange and unexpected things. The upward rush of chimney pots and roofs was checked. My inherent aerobatic training, practised for pleasure these many years, had pushed the control column diagonally forward into the far corner of the cockpit, and kicked maximum rudder hard across in the same direction. Shuddering, the inverted nose lifted until the propeller fanned against the sky. Screwed round by the rudder, the massive aeroplane violently flick-rolled its ten tons. The houses dropped and dwindled as the wings swirled right way up through normal lateral level, and wobbled left and right on the edge of stalling. Then the machine regained balance and climbed steadily to the safety of higher skies.

Moving my head with cautious deliberation, as though it might disturb the machine's precarious equilibrium, I looked through the side windows. Both ailerons were deflected to their maximum upward angle, locked hard against their stops. One must have broken free from the controlling mechanism, and been flung full up by the press of air upon the aerodynamic balance in front of its hinge. The opposite aileron was still serviceable, and only then did I see that it was my own hand which held it firmly applied to counteract the violent cartwheeling effect of the other wing. Maybe five seconds had elapsed from start to finish of that wild gyration—five swift seconds separating unsuspecting sunlit peace from restored and grateful sunlight. Few had escaped from so close a call as this.

Up . . . up . . . gently . . . steadily . . . I willed her towards higher skies, still holding the stick unnaturally across the cockpit to balance the live aileron against the broken. Up and up we laboured towards the life-giving sun, where height spells safety for a parachute descent should the machine prove uncontrollable when I tried to turn and had to bank a wing once more. The fair and familiar landscape was serene in the sunshine as my gaze swept from horizon to horizon. I was a mote in those skies, floating without volition on the tide of time which would be flowing long after the last man had gone to dust.

With infinite caution I began to turn, and knew at once that a landing was within little more than normal limits of judgement. In a five-mile circuit, barely banking lest we rolled over, I edged the Wyvern round into the approach.

I felt a sense of unreality and insecurity as the machine carried me towards the aerodrome. Each second passed with a strange mental clarity. Each threatened that with the next the Wyvern must lurch violently, roll wing down, and drop uncontrollably as it had done three minutes earlier. But cold logic insisted that the flight was safe—that the upturned aileron gave lateral equilibrium far greater than the machine's normal stability. Had not the same thing happened to me years ago when the Whirlwind aileron control burned through?

Nevertheless part of me, crouching, keyed up, waited for the wing to drop; waited for the rapidly approaching moment when the glide would bring the machine below the height at which the cartridge ejector seat offered a last reasonable chance of parachute escape.

I pressed the transmitter switch and gave my call sign. 'Judwin, priority landing. Finals from the west.'

'Judwin, finals,' confirmed a dispassionate voice. But if the voice sounded unfeelingly disinterested, the apparent lack of concern was belied by the sudden movement of fire tender and ambulance towards the point where I should touch down.

The undercarriage locked into position with its usual thump, and without, to my relief, any disturbance of trim sometimes produced by lack of phase in the movements of each leg.

I had plenty of height in hand, and opened the flaps gently, notch by notch, to maximum lift. The right wing sank a fraction under the changed airflow. No aileron could be applied to check the motion. The wing rose level, stabilized by the negative tip angle of the upturned ailerons.

Everything in order: the engine running fast: the aerodrome widening from a finger strip to a broad, green path. The red of the fire tender stood out beyond the trees, its crew looking up at the heavy monoplane. The boundary hedge flashed past, and the tyres touched the turf at 100 m.p.h.

I cut the engine. The turbine slowed with a sharp indrawn breath, followed by the diminishing whisper of the changing pitch of the propeller. My life on earth had begun anew.

Seven times that aeroplane brought me to grave hazard. From the eighth I escaped by the toss of a coin, which decided that Peter Garner, my close friend and assistant chief test pilot, would fly for some air to air photography of the Wyvern.

'I'd like to see my name in the papers for once instead of yours,' he joked.

A final close-up completed the air photography session. Peter gave a friendly wave of farewell, and turned for base.

Within seconds the propeller bearing failed. He tried to

make a forced landing in a very small field, but was unable to flare off the glide. The Wyvern hit the ground with under-carriage up and burst into flames.

Peter and Jimmy, Michael and Ted—yes, and all the others who had died untimely in the preceding years both of peace and war. With what light-hearted defiance they flung their brief days against the weighted scales of time, ventured the unattained, achieved their heart's desire, and in the moment of seeming triumph surrendered the flame of life. These were the terms of the gamble: not an eye for an eye and a tooth for a tooth, but their own irreplaceable lives wagered against the spirit's aspiration—and always in the background was the muffled toll of bells.

Though man is born to trouble as the sparks fly upward, youth does not live with introspective premonition in the sunshine of his life. The consciousness of peril becomes sub-limated in other things, for it seems so safe high in the air with the world remote, and its people and occasions almost for-gotten. In any exploration of the boundaries of knowledge Fate will stab with the same sudden blindness as in everyday normal life. If in all things there are moments when luck alone can bring one through, a pilot's practised judgement has been conditioned by difficult circumstances to be prepared for the abnormal. So the most hazardous incidents of prototype flying leave no greater impression than the many fleeting moments of stress and difficulty encountered during the routine operation of fully developed production aeroplanes. Controls fail, structures break, stability vanishes, loads increase beyond the power of arms and feet, engines catch fire, propellers race and break, jet turbines overspeed or lose their blades, and every prototype I have ever flown has had at least one engine failure resulting in a forced landing. Nevertheless such things are quickly forgotten in the accepted daily risks that are inseparably part of the technique of flying.

Flight must always be made in such a manner that the unknown can be met in a way that leaves some avenue of escape. As a last resort there is the parachute: sometimes fate trembles on the finest thread.

If anything, danger enhanced, rather than detracted from, the peerless fascination of gazing from the heights at the loveliness of the earth: a fascination poignant in the knowledge that such beauty bloomed for every man with the transience of a summer flower. Whatever befell, each flight remained for me a reincarnation. Always there was new discovery of old and familiar things. Some days would be remembered for ever, whilst others might slip into the limbo of forgotten love. Such small things make remembrance: a look, a touch, a sudden vision of perfection, or the darkness of despair. And some gained forgetfulness because they were only part of a mere sequence leading to the unexpected.

So it was with the strike-fighter and the thousand flights I made with her in one variant or another during the course of her protracted trials. Despite difficulties of development greater than the customary share the Wyvern gave hundred-fold enthralment for every hazard. Day after day and many times each day, breathing oxygen automatically controlled, we would climb through the blue skies, until high enough to dive at maximum speed, often pulling out so heavily that iron bands tightened on the temples, our jaws sagged, and the skies darkened as black-out point was reached. Controls applied at extreme angles made the ten-ton fighter lurch and slither in fantastic angles and attitudes as forces and movements were measured to provide the data for long calculations by the technical staff, structural alterations of controls or stabilizing surfaces, and a further succession of check flights. So the development of the aeroplanes continued. Sometimes the imagined remedy made the problem worse, but it was only a temporary setback in the gradual elimination of faults and the evolution of responses which custom had standardized as safe.

From the warm stillness of the bubble-canopied cockpit I watched, in the pauses between each test, the endless permutations of sky and sea and land. Any place in England was within an hour's reach. In a few minutes of full-throttle I could leave the landscape far below, reaching steeply upwards to re-discover the eternities of sun-filled space. At other times I might linger at low altitudes, flaps extended, the turbine whirling with the quiet rumbling of low power, and the propellers spinning their strange shadow of a cross fixedly ahead.

Beneath the slow passage of my wings the pattern of mounds and scars on the downland once more told a moment's story: gleaming chalk-cliffs revealed the epic of ancient waters; and granite bastions defied the menacing seas. By grace of flight England became mine for the possessing. For a few brief minutes I was no longer a pilot flying a creation of complex mechanism, but disembodied, sailing the heavens on angel's wings.

The months of developing the Wyvern mounted into years, until eventually they brought in the late hours of a calm summer afternoon a production machine for test. The haze which for some days had tinted the landscape with tired blue and drawn the horizon close, had given way to crystal clear visibility. On a faint drift of air two miles above the earth an armada of great cumulus slowly sailed. From 20,000 feet above them I could see hundreds of square miles of green and corn-coloured fields stretching into the distance, every line of their blue-green hedges sharp and distinct.

At far intervals in the stillness of the lower skies faint curtains of smoke delineated the shape of distant towns. Demarcations of coast etched the outlines of the lands north and south. On my right was the broad rift of the Bristol Channel, one shore visible from Pembroke far up the Severn into Gloucester. The other embraced Wessex, stretching past Exmoor to Bideford Bay which glittered like a mirror below the shadowed line of the Atlantic. Beyond my left shoulder spread the blue sequin sea of the Channel, its deckled coastline painted with brilliant detail all the way from Beachy Head westward to the dark loom of Start silhouetted above the deep blue of Falmouth Bay.

Time was pressing. Exceptional though the view might be today it must wait for some less hurried moment. It was my possession; and would always be waiting there no less enchanting whenever I needed to attain it. But in this hour I must be occupied with more urgent matters. I accepted the compromise that had been made so many times before. The readings of the rows of dials were noted on my pad. Now for the dive.

The hood-lock was checked, harness straps strained tighter; trimmers re-set. After a quick scrutiny for other aircraft, and a last glance at the instrument panel, I depressed the long nose. The slipstream rose in a crescendo until the limiting speed was attained and held. I crouched in the cabin—head bent to the instruments, senses tuned to the vibrations of the aeroplane, diving on the gyros blind to the outside world. The touch of the controls and the sound of the engine registered that all was well. At 7,000 feet I pulled steadily into level flight and slowed with throttled engine. I glanced outside for a visual check on my position, and I saw again, with a sense of benediction, the far beauty of the waiting landscape. But others too were waiting. I called the dispersal aerodrome and began the brief descent.

Five minutes later I taxied to the tarmac apron by the entrance to the hangar. Cocks were cut and electrics switched off. The engine gasped to a stop and all was silent.

I cranked back the hood, and watched a ladder being dragged to the wing. The foreman clambered up:

'They want you to call at the administration office on your way home,' he reported. 'They said it's urgent.'

I nodded absently and unpinned my safety harness and parachute.

'That elevator was much better,' I said. 'We'll do the aft loading tomorrow.'

Through the winding lanes and roads of a dreamy countryside the car carried me swiftly back to the offices. I took the photographic results to the technical section and turned to the administration block. It was not until I went up the stairs that I wondered what could be so urgent.

In kindly fashion they broke the news: 'There's another job we'd like you to tackle. We think it is time you gave up test-flying. Twenty-five years is a longish spell.'

What were they telling me? Give up flying? Twenty-five years? . . . It was long before that I had started to fly. . . . With no warning was this the end? Why had there been no sign, no portent in the skies—only an urgent test that gave no time for the leisured enjoyment of flight?

The news was like the sudden catastrophe which in those years of flying had many times disrupted the oblivious quiet. But just as a sleeping awareness of danger had underlain my every flight, so I had always known that one day there must come an end to this enchanting quest of the skies. Better suddenly like this than shirking from the last farewell, the final touch, and the emptiness as one turns away.

Yet it seemed impossible to say 'Good-bye.'

Year after year the skies had been mine—the eternal skies, magically repainted from hour to hour with changing combinations of mist and cloud and light. So often in that vast blue emptiness I had sat absorbed and silent, enfolded in my thoughts and action, waiting without impatience for time to go. If now I could but gather it up again and fly through time once more: the space, the freedom, the happiness, the answer almost heard.

'We know it will be a wrench—but let's have your decision tomorrow,' they said.

In the calm evening I slowly walked home through the woods. No tree stirred. I walked without consciousness. No bird sang—for the time of their rapture was gone. I looked up. In the highest zenith a contrail grew across the cloudless blue— but there was no echo in the sky. Everything was silent. The trail travelled swiftly to the far horizon, the aeroplane unseen, leaving only the slowly expanding vapour path to tell of the quietness that in the infinities of outer space awaits the quenchless, questing spirit of man.

Literature and History of Aviation

AN ARNO PRESS COLLECTION

Arnold, H[enry] H.
Global Mission. 1949.

Bordeaux, Henry.
Georges Guynemer: Knight of the Air. Translated by Louise Morgan Sill. 1918.

Boyington, "Pappy" (Col. Gregory Boyington).
Baa Baa Black Sheep. 1958.

Buckley, Harold.
Squadron 95. 1933.

Caidin, Martin.
Golden Wings. 1960.

"Contact" (Capt. Alan Bott).
Cavalry of the Clouds. 1917.

Crossfield, A. Scott and Clay Blair, Jr.
Always Another Dawn. 1960.

Fokker, Anthony H. G. and Bruce Gould.
Flying Dutchman: The Life of Anthony Fokker. 1931.

Gibson, Guy.
Enemy Coast Ahead. 1946.

Goldberg, Alfred, editor.
A History of the United States Air Force 1907-1957. 1957.

Gurney, Gene.
Five Down and Glory. Edited by Mark P. Friedlander, Jr. 1958.

Hall, Norman S.
The Balloon Buster: Frank Luke of Arizona. 1928.

Josephson, Matthew.
Empire of the Air: Juan Trippe and the Struggle for World Airways. 1944.

Kelly, Charles J., Jr.
The Sky's the Limit: The History of the Airlines. 1963.
New Introduction by Charles J. Kelly, Jr.

Kelly, Fred C., editor.
Miracle at Kitty Hawk. 1951.

La Farge, Oliver.
The Eagle in the Egg. 1949.

Levine, Isaac Don.
Mitchell: Pioneer of Air Power. 1943.

Lougheed, Victor.
Vehicles of the Air. 1909.

McFarland, Marvin W., editor.
The Papers of Wilbur and Orville Wright. 2 volumes. 1953.

McKee, Alexander.
Strike From the Sky: The Story of the Battle of Britain. 1960.

Macmillan, Norman.
Into the Blue. 1969.

Magoun, F. Alexander and Eric Hodgins.
A History of Aircraft. 1931.

Parsons, Edwin C.
I Flew with the Lafayette Escadrille. 1963.

Penrose, Harald.
No Echo in the Sky. 1958.

Reynolds, Quentin.
The Amazing Mr. Doolittle. 1953.

Saunders, Hilary St. George.
Per Ardua: The Rise of British Air Power 1911-1939. 1945.

Stilwell, Hart and Slats Rodgers.
Old Soggy No. 1. 1954.

Studer, Clara.
Sky Storming Yankee: The Life of Glenn Curtiss. 1937.

Turnbull, Archibald D. and Clifford L. Lord.
History of United States Naval Aviation. 1949.

Turner, C. C.
The Old Flying Days. 1927.

Von Richthofen, Manfred F.
The Red Air Fighter. 1918.

Werner, Johannes.
Knight of Germany: Oswald Boelcke, German Ace. Translated by
  Claud W. Sykes. 1933.

Wise, John.
Through the Air. 1873.

Wolff, Leon.
Low Level Mission. 1957.

Yakovlev, Alexander.
Notes of an Aircraft Designer. Translated by Albert Zdornykh. n.d.